A Guide to NIP Theories

The study of NIP theories has received much attention from model theorists in the last decade, fuelled by applications to o-minimal structures and valued fields. This book, the first to be written on NIP theories, is an introduction to the subject that will appeal to anyone interested in model theory: graduate students and researchers in the field, as well as those in nearby areas such as combinatorics and algebraic geometry. Without dwelling on any one particular topic, it covers all of the basic notions and gives the reader the tools needed to pursue research in this area. An effort has been made in each chapter to give a concise and elegant path to the main results and to stress the most useful ideas. Particular emphasis is put on honest definitions, handling of indiscernible sequences and measures. The relevant material from other fields of mathematics is made accessible to the logician.

PIERRE SIMON is Chargé de recherche, CNRS at Université Lyon 1 in France. He completed his PhD at Université Paris-Sud, Orsay under the supervision of Elisabeth Bourscaren. His thesis, "Ordre et stabilité dans les théories NIP", received the 2012 Sacks Prize for the best thesis in logic that year as well as the Perrissin-Pirasset/Schneider prize from the Chancellerie des Universités de Paris.

LECTURE NOTES IN LOGIC

A Publication for
The Association for Symbolic Logic

This series serves researchers, teachers, and students in the field of symbolic logic, broadly interpreted. The aim of the series is to bring publications to the logic community with the least possible delay and to provide rapid dissemination of the latest research. Scientific quality is the overriding criterion by which submissions are evaluated.

More information, including a list of the books in the series, can be found at http://www.aslonline.org/books-lnl.html

A Guide to NIP Theories

PIERRE SIMON
CNRS, Université Lyon 1

ASSOCIATION FOR SYMBOLIC LOGIC

CAMBRIDGE
UNIVERSITY PRESS

CAMBRIDGE
UNIVERSITY PRESS

University Printing House, Cambridge CB2 8BS, United Kingdom

Cambridge University Press is part of the University of Cambridge.

It furthers the University's mission by disseminating knowledge in the pursuit of education, learning and research at the highest international levels of excellence.

www.cambridge.org
Information on this title: www.cambridge.org/9781107057753

First published 2015

A catalogue record for this publication is available from the British Library

Library of Congress Cataloging-in-Publication Data
Simon, Pierre, 1985–
A guide to NIP theories / Pierre Simon, Université Lyon I.
pages cm. – (Lecture notes in logic)
Includes bibliographical references and index.
ISBN 978-1-107-05775-3 (Hardback : alk. paper)
1. Model theory. 2. Independence (Mathematics). 3. Logic, Symbolic and mathematical. I. Title.
QA9.7.S536 2015
511.3′4–dc23 2015013315

ISBN 978-1-107-05775-3 Hardback

CONTENTS

Chapter 1

INTRODUCTION

This text is an introduction to the study of NIP (or dependent) theories. It is meant to serve two purposes. The first is to present various aspects of NIP theories and give the reader sufficient background material to understand the current research in the area. The second is to advertise the use of honest definitions, in particular in establishing basic results, such as the so-called *shrinking of indiscernibles*. Thus although we claim no originality for the theorems presented here, a few proofs are new, mainly in chapters 3, 4 and 9.

We have tried to give a *horizontal* exposition, covering different, sometimes unrelated topics at the expense of exhaustivity. Thus no particular subject is dealt with in depth and mainly *low-level* results are included. The choices made reflect our own interests and are certainly very subjective. In particular, we say very little about algebraic structures and concentrate on combinatorial aspects. Overall, the style is concise, but hopefully all details of the proofs are given. A small number of facts are left to the reader as exercises, but only once or twice are they used later in the text.

The material included is based on the work of a number of model theorists. Credits are usually not given alongside each theorem, but are recorded at the end of the chapter along with pointers to additional topics.

We have included almost no preliminaries about model theory, thus we assume some familiarity with basic notions, in particular concerning compactness, indiscernible sequences and ordinary imaginaries. Those prerequisites are exposed in various books such as that of Poizat [95], Marker [81], Hodges [56] or the recent book [115] by Tent and Ziegler. The material covered in a one-semester course on model theory should suffice. No familiarity with stability theory is required.

History of the subject. In his early works on classification theory, Shelah structured the landscape of first order theories by drawing dividing lines defined by the presence or absence of different combinatorial configurations. The most important one is that of stability. In fact, for some twenty years, pure model theory did not venture much outside of stable theories. Shelah discovered the independence property when studying the possible behaviors for the function relating the size of a subset to the number of types over it. The class of theories lacking the independence property, or NIP theories, was studied very little in the early days. However some basic results were established, mainly by Shelah and Poizat (see [95, Chapter 12] for an account of those works).

As the years passed, various structures were identified as being NIP: most notably, Henselian valued fields of characteristic 0 with NIP residue field and ordered group (Delon [32]), the field $\mathbb{Q}_p$ of p-adics (see Bélair [15]) and ordered abelian groups (Gurevich and Schmitt [50]). However, NIP theories were not studied *per se*. In [91], Pillay and Steinhorn, building on work of van den Dries, defined o-minimal theories as a framework for tame geometry. This has been a very active area of research ever since. Although it was noticed from the start that o-minimal theories lacked the independence property, very little use of this fact was made until recently. Nevertheless, o-minimal theories provide a wealth of interesting examples of NIP structures.

In the years since 2000, the interest in NIP theories has been rekindled and the subject has been expanding ever since. First Shelah initiated a systematic study which lead to a series of papers: [107], [109], [102], [111], [110]. Amongst other things, he established the basic properties of forking, generalized a theorem of Baisalov and Poizat on externally definable sets, defined some subclasses, so called "strongly dependent" and "strongly$^+$ dependent". This work culminates in [110] with the proof that NIP theories have few types up to automorphism (over saturated models). Parallel to this work, Hrushovski, Peterzil and Pillay developed the theory of *measures* (a notion introduced by Keisler in [71]) in order to solve Pillay's conjecture on definably compact groups in o-minimal theories.

A third line of research starts with the work of Hrushovski, Haskell and Macpherson on algebraically closed valued fields (ACVF) and in particular on *metastability* ([52]). This lead to Hrushovski and Loeser giving a model theoretic construction of Berkovich spaces in rigid geometry as spaces of stably-dominated types, which made an explicit use of the NIP property along with the work on metastability.

Motivated by those results, a number of model theorists became interested in the subject and investigated NIP theories in various directions. We will present some in the course of this text and mention others at the end of each chapter. It is not completely clear at this point how the subject will develop and what topics will turn out to be the most fruitful.

Let us end this general introduction by mentioning where NIP sits with respect to other classes of theories. First, all stable theories are NIP, as are o-minimal and C-minimal theories. Another well-studied extension of stability is that of simple theories (see Wagner [122]), however it is in a sense orthogonal to NIP: a theory is both simple and NIP if and only if it is stable. Simple and NIP theories both belong to the wider class of NTP_2 theories (defined in Chapter 5).

Organization of this text. Aside from the introduction and appendices, the text is divided into 8 chapters, each one focussing on a specific topic. In Chapter 2, we present the classical theory as it was established by Shelah and Poizat. We first work formula-by-formula giving some equivalent definitions of NIP. We then move to invariant types and Morley sequences. Starting then, and throughout most of the text, we assume that our ambient theory T is NIP. That assumption will be dropped only for the first three sections of Chapter 5. Many results could be established for an NIP *formula* (or type) inside a (possibly) independent theory, but for the sake of clarity we will not work at this level of generality.

The end of Chapter 2 is a collection of appendices on different subjects. We study dense trees in some detail as one can obtain from them a lot of intuition on NIP theories, we recall basic facts on stable theories, discuss the strict order property and give the original characterization of NIP theories by counting types.

In Chapter 3 we define honest definitions. They serve as a substitute to definability of types in NIP theories. We use them to prove Shelah's theorem on expanding a model by externally definable sets and the very important results about shrinking of indiscernibles.

Chapter 4 deals with dp-rank and strong dependence. In the literature, one can find up to three different definitions of dp-rank, based on how one handles the problem of *almost finite*, non-finite rank. None of them is perfect, but we have decided to use the same convention as in Adler's paper [2] on burden, although we refrain from duplicating limit cardinals into κ_- and κ. Instead, we define when dp-rk$(p) < \kappa$, and it may happen that the dp-rank of a type is not defined (for example it can be $< \aleph_0$ but greater than all integers).

In Chapter 5, we study forking and dividing. The main results are $bdd(A)$-invariance of non-forking extensions (Hrushovski and Pillay [61]) and equality of forking and dividing over models (Chernikov and Kaplan [25]). The right context for this latter result is NTP_2 theories, but here again we assume NIP which slightly simplifies some proofs.

The next three chapters have a different flavor. In Chapter 6, we change the framework to that of finite combinatorics. We are concerned with families of finite VC-dimension over finite sets. The finite and infinite approaches come together to prove uniformity of honest definitions. In Chapter 7, the two frameworks are combined with the introduction of Keisler measures. The most important class of examples of such measures is that of translation-invariant measures on definable groups. Those are investigated in Chapter 8. We also discuss there connected components of groups.

The last chapter addresses the problem of characterizing NIP structures which are in some sense completely unstable. They are called *distal* structures.

Finally, two appendices are included. The first one gives some algebraic examples and in particular records some facts about valued fields. Most of the proofs are omitted, but we explain how to show that those structures are NIP. The other appendix is very short and collects results about probability theory for reference in the text.

Acknowledgments. Part of the material presented here was exposed in 2011 during a series of lectures in Paris. I would like to thank all the participants of this seminar: Élisabeth Bouscaren, Zoé Chatzidakis, Pablo Cubides Kovacsics, Françoise Delon, Martin Hils, Samaria Montenegro, Françoise Point, Silvain Rideau and Patrick Simonetta. I have received very helpful advice from a number of people: David Bradley-Williams, Rafel Farré, Martin Hils, Udi Hrushovski, Itay Kaplan, Dugald Macpherson, Dave Marker and the two anonymous referees. Special thanks to Alex Kruckman for his extensive list of comments and corrections.

1.1. Preliminaries

We work with a complete, usually one-sorted, theory T in a language L. We have a monster model $\mathcal{U}$ which is $\bar{\kappa}$-saturated and homogeneous. A subset $A \subset \mathcal{U}$ is *small* if it is of size less than $\bar{\kappa}$. For $A \subseteq \mathcal{U}$, $L(A)$ denotes the set of formulas with parameters in A. In particular, $\phi(x) \in L$ means that ϕ is without parameters.

We do not usually distinguish between points and tuples. If a is a tuple of size $|a|$, we will write $a \in A$ to mean $a \in A^{|a|}$. Similarly, letters such as $x, y, z, \ldots$ are used to denote tuples of variables.

We often work with *partitioned formulas*, namely formulas $\phi(x; y)$ with a separation of variables into object and parameters variables. The intended partition is indicated with a semicolon.

If $A \subset \mathcal{U}$ is any set, and $\phi(x)$ is a formula, then $\phi(A) = \{a \in A^{|x|} : \mathcal{U} \models \phi(a)\}$. The set of types over A in the variable x is denoted by $S_x(A)$. We will often drop the x. If $p \in S_x(A)$, we might write p_x or $p(x)$ to emphasize that p is a type in the variable x. We say that a type p *concentrates* on a definable set $\phi(x)$ if $p \vdash \phi(x)$. If $A \subseteq B$ and p is a type over B, we denote by $p \upharpoonright A$, or by $p|_A$ the *restriction* of p to A.

We will often write either $\models \phi(a)$ or $a \models \phi(x)$ to mean $\mathcal{U} \models \phi(a)$, and similarly for types.

We use the notation ϕ^0 to mean $\neg\phi$ and ϕ^1 to mean ϕ. If $\phi(x; y)$ is a partitioned formula, a *ϕ-type* over A is a maximal consistent set of formulas of the form $\phi(x; a)^\varepsilon$, for $a \in A$ and $\varepsilon \in \{0, 1\}$. The set of ϕ-types over A is denoted by $S_\phi(A)$.

A *global type*, is a type over $\mathcal{U}$.

The group of *automorphisms* of $\mathcal{U}$ is denoted by $\mathrm{Aut}(\mathcal{U})$, whereas $\mathrm{Aut}(\mathcal{U}/A)$ refers to the subgroup of $\mathrm{Aut}(\mathcal{U})$ of automorphisms fixing A pointwise.

1.1.1. Indiscernible sequences. We will typically denote sequences of tuples by $I = (a_i : i \in \mathfrak{I})$ where $\mathfrak{I}$ is some linearly ordered set. The order on $\mathfrak{I}$ will be denoted by $<_{\mathfrak{I}}$ or simply $<$ if no confusion arises. If $I = (a_i : i \in \mathfrak{I})$ and $J = (b_j : j \in \mathfrak{J})$, then we write the concatenation of I and J as $I + J$. It has I as initial segment and J as the complementary final segment. We use the notation (a) to denote the sequence which has a as unique element.

We say that the sequence I is *endless* if the indexing order $\mathfrak{I}$ has no last element.

Let Δ be a finite set of formulas and A a set of parameters. A (possibly finite) sequence $I = (a_i : i \in \mathfrak{I})$ is Δ-indiscernible over A, if for every integer k and two increasing tuples $i_1 <_{\mathfrak{I}} \cdots <_{\mathfrak{I}} i_k$ and $j_1 <_{\mathfrak{I}} \cdots <_{\mathfrak{I}} j_k$, $b \in A$ and formula $\phi(x_1, \ldots, x_k; y) \in \Delta$, we have $\phi(a_{i_1}, \ldots, a_{i_k}; b) \leftrightarrow \phi(a_{j_1}, \ldots, a_{j_k}; b)$. An *indiscernible sequence* is an *infinite* sequence which is Δ-indiscernible for all Δ.

Let $I = (a_i : i \in \mathfrak{I})$ be any sequence. We define the Ehrenfeucht-Mostowski type (or *EM-type*) of I over A to be the set of $L(A)$-formulas $\phi(x_1, \ldots, x_n)$ such that $\mathcal{U} \models \phi(a_{i_1}, \ldots, a_{i_n})$ for all $i_1 < \cdots < i_n \in \mathfrak{I}$, $n < \omega$. If I is an indiscernible sequence, then for every n, the restriction of the EM-type of I to formulas in n variables is a complete type over A. If $A = \emptyset$, then we can omit it. We will write $I \equiv_A^{EM} J$ to mean that I and J are two A-indiscernible sequences having the same EM-type over A. If I is any sequence and $\mathfrak{J}$ is any infinite linear order, then using Ramsey's theorem and compactness, we can find an indiscernible sequence J indexed by $\mathfrak{J}$ and realizing the EM-type of I (see e.g., [115, Lemma 5.1.3]).

A sequence I is *totally indiscernible* (or set indiscernible) if every permutation of it is indiscernible. If a sequence $(a_i : i \in \mathfrak{I})$ is not totally indiscernible, then there is some formula $\phi(x, y)$, possibly with parameters, which orders it, that is such that $\phi(a_i, a_j)$ holds if and only if $i \leq j$.

Most of the time, Ramsey and compactness will be sufficient for us to construct indiscernible sequences. However, we will need once or twice a more powerful result which is an easy application of the Erdős-Rado theorem.

PROPOSITION 1.1. *Let A be a set of parameters, $\kappa > |T| + |A|$ and $\lambda = \beth_{(2^\kappa)^+}$. Let $(a_i : i < \lambda)$ be a sequence of tuples all of the same size $\leq \kappa$. Then there is an indiscernible sequence $(b_i : i < \omega)$ such that for any $i_1 < \cdots < i_n < \omega$, there are some $j_1 < \cdots < j_n < \lambda$ with*

$$a_{i_1} \ldots a_{i_n} \equiv_A b_{j_1} \ldots b_{j_n}.$$

See e.g. [21, Proposition 1.6] for a proof.

Chapter 2

THE NIP PROPERTY AND INVARIANT TYPES

In this chapter, we introduce the basic objects of our study. We first define the notion of an NIP formula. The combinatorial definition is not very handy, and we give an equivalent characterization involving indiscernible sequences which is the one we will most often use. We then define NIP theories as theories in which all formulas are NIP and give some examples. We discuss invariant types and their relation to indiscernible sequences. In particular, we define generically stable types, which share some characteristics with types in stable theories.

To illustrate the notions considered, we prove some results on definable groups in NIP theories: the Baldwin-Saxl theorem, and Shelah's theorem on existence of definable envelopes for abelian subgroups.

In the "additional topics" section we introduce trees, which serve as a paradigm for NIP theories. Many examples of NIP theories are either explicitly constructed as a tree with additional structure, or have an underlining tree-structure (valued fields for example). We discuss in more details the theory of dense meet-trees, and in particular describe indiscernible sequences in it. The next subsection collects some facts about stable formulas and theories. We then present the strict order property and finally give yet another characterization of NIP in terms of counting types.

2.1. NIP formulas

Let $\phi(x; y)$ be a partitioned formula. We say that a set A of $|x|$-tuples is *shattered* by $\phi(x; y)$ if we can find a family $(b_I : I \subseteq A)$ of $|y|$-tuples such that

$$\mathcal{U} \models \phi(a; b_I) \iff a \in I, \quad \text{for all } a \in A.$$

By compactness, this is equivalent to saying that every finite subset of A is shattered by $\phi(x; y)$.

DEFINITION 2.1. A partitioned formula $\phi(x; y)$ is NIP (or dependent) if no infinite set of $|x|$-tuples is shattered by $\phi(x; y)$.

If a formula is not NIP, we say that it has IP.

Remark 2.2. The acronym IP stands for the *Independence Property* and NIP is its negation. Some authors (notably Shelah) use the terminology dependent/independent instead of NIP/IP.

Remark 2.3. If $\phi(x; y)$ is NIP, then by compactness, there is some integer n such that no set of size n is shattered by $\phi(x; y)$.

The maximal integer n for which there is some A of size n shattered by $\phi(x; y)$ is called the *VC-dimension* of ϕ. If there is no such integer, that is if the formula ϕ has IP, then we say that its VC-dimension is infinite.

Example 2.4.

- Let T be DLO: the theory of dense linear orders with no endpoints. Then the formula $\phi(x; y) = (x \leq y)$ is NIP of VC-dimension 1. Indeed, if we have $a_1 < a_2$, then we cannot find some $b_{\{2\}}$ such that
$$\mathcal{U} \models \neg\phi(a_1; b_{\{2\}}) \wedge \phi(a_2; b_{\{2\}}).$$
- If $\phi(x; y)$ is a stable formula, then it is NIP (see Section 2.3.2 if needed).
- If T is the theory of arithmetic, then the formula $\phi(x; y) =$ "x divides y" has IP. To see this, take any $N \in \mathbb{N}$ and $A = \{p_0, \dots, p_{N-1}\}$ a set of distinct prime numbers. For any $I \subseteq N$, set b_I to be $\prod_{i \in I} p_i$. We have $\models \phi(p_i, b_I) \iff i \in I$. Thus the set A is shattered and $\phi(x; y)$ has infinite VC-dimension.
- If T is the random graph in the language $L = \{R\}$, then the formula $\phi(x; y) = xRy$ has IP. In fact any set of elements is shattered by ϕ.
- If T is a theory of an infinite Boolean algebra, in the natural language $\{0, 1, \neg, \vee, \wedge\}$, then the formula $x \leq y$ (defined as $x \wedge y = x$) has IP. Indeed, it shatters any set A with $a \wedge b = 0$ for $a \neq b \in A$.

If $\phi(x; y)$ is a partitioned formula, we let $\phi^{opp}(y; x) = \phi(x; y)$. Hence ϕ^{opp} is the same formula as ϕ, but we have exchanged the role of variables and parameters. The following fact will be used throughout this text, often with no explicit mention.

LEMMA 2.5. *The formula $\phi(x; y)$ is NIP if and only if $\phi^{opp}(y; x)$ is NIP.*

PROOF. Assume that $\phi(x; y)$ has IP. Then by compactness, we can find some $A = \{a_i : i \in \mathfrak{P}(\omega)\}$ which is shattered by $\phi(x; y)$ as witnessed by tuples b_I, $I \subseteq \mathfrak{P}(\omega)$. Let $B = \{b_j : j \in \omega\}$ where $b_j := b_{I_j}$ and $I_j := \{X \subseteq \omega : j \in X\}$. Then for any $J_0 \subseteq \omega$, we have

$$\models \phi(a_{J_0}, b_j) \iff j \in J_0.$$

This shows that B is shattered by ϕ^{opp}. Therefore ϕ^{opp} has IP. ⊣

Remark 2.6. The VC-dimension of a formula ϕ need not be equal to the VC-dimension of the opposite formula ϕ^{opp}. For example, let T be the theory of equality, then the formula $\phi(x; y_1y_2y_3) = (x = y_1 \vee x = y_2 \vee x = y_3)$ has VC-dimension 3, but the opposite formula only has VC-dimension 2.

See Lemma 6.3 for inequalities linking the VC-dimensions of ϕ and ϕ^{opp}. For now, our only concern is whether they are finite or not.

We now give an equivalent characterization of NIP which is often the most convenient one to use.

LEMMA 2.7. *The formula $\phi(x; y)$ has IP if and only if there is an indiscernible sequence $(a_i : i < \omega)$ and a tuple b such that*

$$\models \phi(a_i; b) \iff i \text{ is even}.$$

PROOF. ($\Leftarrow$): Assume that there is a sequence $(a_i : i < \omega)$ and a tuple b as above. Let $I \subseteq \omega$. We show that there is some b_I such that $\phi(a_i; b_I)$ holds if and only if $i \in I$. We can find an increasing one-to-one map $\tau : \omega \to \omega$ such that for all $i \in \omega$, $\tau(i)$ is even if and only if i is in I. Then by indiscernibility the map sending a_i to $a_{\tau(i)}$ for all $i < \omega$ is a partial isomorphism. It extends to a global automorphism σ. Then take $b_I = \sigma^{-1}(b)$.

($\Rightarrow$): Assume that $\phi(x; y)$ has IP. Let $A = (a_i : i < \omega)$ be a sequence of $|x|$-tuples which is shattered by $\phi(x; y)$. By Ramsey and compactness, we can find some indiscernible sequence $I = (c_i : i < \omega)$ of $|x|$-tuples realizing the EM-type of A. It follows that for any two disjoint finite sets I_0 and I_1 of I, the partial type $\{\phi(x; c) : c \in I_0\} \cup \{\neg\phi(x; c) : c \in I_1\}$ is consistent. Then by compactness, I is shattered by $\phi(x; y)$. In particular, there is b such that $\phi(c_i; b)$ holds if and only if i is even. ⊣

Let $\phi(x; y)$ be an NIP formula, then there is a finite set Δ of formulas and an integer $n_{\phi,\Delta}$ such that the following do not exist:

- $(a_i : i < n_{\phi,\Delta})$ a Δ-indiscernible sequence of $|x|$-tuples;
- b a $|y|$-tuple, such that $\neg(\phi(a_i; b) \leftrightarrow \phi(a_{i+1}; b))$ holds for $i < n_{\phi,\Delta} - 1$.

Indeed, if we could not find such Δ and $n_{\phi,\Delta}$, then the partial type in variables $(x_i : i < \omega)^\frown y$ stating that $(x_i : i < \omega)$ is an indiscernible sequence and $\neg(\phi(x_i; y) \leftrightarrow \phi(x_{i+1}; y))$ holds for all $i < \omega$ would be consistent, contradicting the previous lemma.

Let $I = (a_i : i \in \mathcal{I})$ be an indiscernible sequence and take an NIP formula $\phi(x; y) \in L$ and a tuple of parameters $b \in \mathcal{U}$. Then there is a maximal integer n such that we can find $i_0 < \cdots <_{\mathcal{I}} i_n$ with $\neg(\phi(a_{i_k}; b) \leftrightarrow \phi(a_{i_{k+1}}; b))$ for all $k < n$. We call such an n the number of alternations of $\phi(x; b)$ on the sequence I and write it as $\mathrm{alt}(\phi(x; b), I)$. We let $\mathrm{alt}(\phi(x; y))$ denote the maximum value of $\mathrm{alt}(\phi(x; b), I)$ for b ranging in $\mathcal{U}$ and I ranging over all indiscernible sequences. Note that this maximum exists and is bounded by the number $n_{\phi,\Delta}$ of the previous paragraph. We sometimes call $\mathrm{alt}(\phi(x; y))$ the *alternation rank* (or *number*) of $\phi(x; y)$.

PROPOSITION 2.8. *The formula* $\phi(x;y)$ *is NIP if and only if for any indiscernible sequence* $(a_i : i \in \mathfrak{I})$ *and tuple* b, *there is some end segment* $\mathfrak{I}_0 \subseteq \mathfrak{I}$ *and* $\varepsilon \in \{0,1\}$ *such that* $\phi(a_i;b)^\varepsilon$ *holds for any* $i \in \mathfrak{I}_0$.

PROOF. If $\mathfrak{I}$ has a last element i_0, let $\mathfrak{I}_0 = \{i_0\}$. Otherwise, this follows immediately from the discussion above. ⊣

Of course the equivalence also holds if we restrict to sequences indexed by ω, or in fact by any given linear order with no last element.

LEMMA 2.9. *A Boolean combination of NIP formulas is NIP.*

PROOF. It is clear from the definition that the negation of an NIP formula is NIP.

Let $\phi(x;y)$ and $\psi(x;y)$ be two NIP formulas and we want to show that $\theta(x;y) = \phi(x;y) \wedge \psi(x;y)$ is NIP. We use the criterion from Proposition 2.8. Let $(a_i : i \in \mathfrak{I})$ be an indiscernible sequence of $|x|$-tuples and let b be a $|y|$-tuple. Let $\mathfrak{I}_\phi \subseteq \mathfrak{I}$ be an end segment such that $\phi(a_i;b) \leftrightarrow \phi(a_j;b)$ holds for $i,j \in \mathfrak{I}_\phi$. Define $\mathfrak{I}_\psi$ similarly and let $\mathfrak{I}_0 = \mathfrak{I}_\phi \cap \mathfrak{I}_\psi$. Then $\mathfrak{I}_0$ is an end segment of $\mathfrak{I}$ and we have $\theta(a_i;b) \leftrightarrow \theta(a_j;b)$ for $i,j \in \mathfrak{I}_0$. This shows that $\theta(x;y)$ is NIP. ⊣

2.1.1. NIP theories.

DEFINITION 2.10. The theory T is NIP if all formulas $\phi(x;y) \in L$ are NIP.

Note that if T is NIP, then also all formulas $\phi(x;y)$ with parameters are NIP, since if $\phi(x;y,d)$ has IP, then so does $\phi(x;y^\frown z)$.

PROPOSITION 2.11. *Assume that all formulas* $\phi(x;y) \in L$ *with* $|y| = 1$ *are NIP, then* T *is NIP.*

PROOF. Assume that all formulas $\phi(x;y)$ with $|y| = 1$ are NIP.

CLAIM. *Let* $(a_i : i < |T|^+)$ *be an indiscernible sequence of tuples, and let* $b \in \mathcal{U}$, $|b| = 1$. *Then there is some* $\alpha < |T|^+$ *such that the sequence* $(a_i : \alpha < i < |T|^+)$ *is indiscernible over* b.

PROOF OF CLAIM. If this does not hold, then for every $\alpha < |T|^+$, for some formula $\delta_\alpha(x_1,\dots,x_{k(\alpha)};y)$, we can find

$$\alpha < i_1 < \cdots < i_{k(\alpha)} < |T|^+ \quad \text{and} \quad \alpha < j_1 < \cdots < j_{k(\alpha)} < |T|^+$$

such that $\models \delta_\alpha(a_{i_1},\dots,a_{i_{k(\alpha)}};b) \wedge \neg\delta_\alpha(a_{j_1},\dots,a_{j_{k(\alpha)}};b)$. There is some formula $\delta(x_1,\dots,x_k;y)$ such that $\delta_\alpha = \delta$ for cofinally many values of α. Then we can construct inductively a sequence $I = (i_1^{l\frown}\dots{}^\frown i_k^l : l < \omega)$ such that $i_1^l < \cdots < i_k^l < i_1^{l+1}$ for all $l < \omega$ and $\delta(a_{i_1^l},\dots,a_{i_k^l},b)$ holds if and only if l is even. As the sequence $(a_{i_1^l}\dots a_{i_k^l} : l < \omega)$ is indiscernible, this contradicts the assumption that $\delta(x_1,\dots,x_k;y)$ is NIP. ⊣ (Claim)

Now let $\phi(x;y)$ be any formula, where $y = y_1{}^\frown\dots{}^\frown y_n$ is an n-tuple. Let $(a_i : i < |T|^+)$ be any indiscernible sequence of $|x|$-tuples and let $b = b_1{}^\frown\dots{}^\frown b_n$ be an n-tuple. By the claim, there is some $\alpha_1 < |T|^+$ such that the sequence

$(a_i : \alpha_1 < i < |T|^+)$ is indiscernible over b_1. This implies that the sequence $(a_i\hat{\ }b_1 : \alpha_1 < i < |T|^+)$ is indiscernible. Therefore another application of the claim gives some $\alpha_1 < \alpha_2 < |T|^+$ such that the sequence $(a_i\hat{\ }b_1 : \alpha_2 < i < |T|^+)$ is indiscernible over b_2. Iterating, we find $\alpha_n < |T|^+$ such that $(a_1\hat{\ }b_1\hat{\ }\dots\hat{\ }b_n : \alpha_n < i < |T|^+)$ is indiscernible. This implies that the truth value of $\phi(a_i; b)$ is constant for $i > \alpha_n$. By Proposition 2.8, the formula $\phi(x; y)$ is NIP. ⊣

Example 2.12. We now give some examples of NIP theories and refer to other sections, in particular the Appendix, for details.

- Any stable theory is NIP. [See Section 2.3.2.]
- Any o-minimal theory is NIP.
  [Recall that a one-sorted theory T is o-minimal if there is a definable total order $\leq$ on the universe such that any definable set in dimension 1 is a finite Boolean combinations of intervals and points. To prove that T is NIP, we use Proposition 2.11 and symmetry to reduce to the case of formulas $\phi(x; y)$, $|x| = 1$. By o-minimality and compactness, there is some n such that any definable set $\phi(x; b)$ is a Boolean combination of up to n intervals and points. Then by Lemma 2.9, it is enough to treat the case of a formula of the form $x \leq y$. But we have already seen that such a formula has VC-dimension 1. See also Section A.1.3.]
- Any theory of a pure linear order or of a pure tree—in particular T_{dt} which is described in Section 2.3.1—is NIP. [Section A.1.2]
- Any C-minimal theory is NIP. The most important example of such is the theory ACVF of algebraically closed valued fields. [Section A.1.4]
- The theory $Th(\mathbb{Q}_p)$ of the p-adics is NIP (either in the pure field language or in the valued field language). [Section A.2]

As one can see from this list, most known NIP theories are either stable or somehow built around a linear order or a tree. We do not know if this is a general fact. A famous open question in this direction is whether any unstable NIP theory interprets an infinite linear order.

To finish this section, we give an example where the NIP property is used to deduce results about definable groups.

THEOREM 2.13 (Baldwin-Saxl). *Let G be a group definable in an NIP theory T. Let H_a be a uniformly definable family of subgroups of G. Then there is an integer N such that for any finite intersection $\bigcap_{a \in A} H_a$, there is a subset $A_0 \subseteq A$ of size N with $\bigcap_{a \in A} H_a = \bigcap_{a \in A_0} H_a$.*

PROOF. The fact that H_a is a uniformly definable family of subgroups means that there is a formula $\phi(x; y)$ such that each H_a is defined by $\phi(x; a)$. Without loss, any instance $\phi(x; a')$ of ϕ defines a subgroup of G.

Take any integer N and assume that the conclusion of the theorem does not hold for N. Then we can find some set $A = \{a_0, \dots, a_N\}$ of parameters such that for every $k \leq N$, we have $\bigcap_{a \in A \setminus \{a_k\}} H_a \neq \bigcap_{a \in A} H_a$. Let $K_k = \bigcap_{a \in A \setminus \{a_k\}} H_a$ and $K = \bigcap_{a \in A} H_a$. For every $k \leq N$, pick a point $c_k \in K_k \setminus K$. For $B \subseteq N + 1$, define $c_B = \prod_{k \in B} c_k$, where the product is in the sense of the group G. Then we have

$$c_B \in H_{a_k} \iff k \notin B.$$

This shows that the formula $\phi^{opp}(y; x) := \phi(x; y)$ has VC-dimension at least N. Therefore there is a maximal such N. ⊣

This theorem will be used in Chapter 8 to prove the existence of the connected component G^0 in NIP groups.

Exercise 2.14. Show that if T is NIP, then so is T^{eq}.

2.2. Invariant types

DEFINITION 2.15 (Invariant type). Let $A \subset \mathcal{U}$ be a small subset and $p \in S_x(\mathcal{U})$. We say that p is A-invariant if $\sigma p = p$ for any $\sigma \in \text{Aut}(\mathcal{U}/A)$.

We say that p is invariant, if it is A-invariant for some small $A \subset \mathcal{U}$.

Another way to phrase the definition is to say that p is A-invariant if for every formula $\phi(x; y)$ and tuples $b, b' \in \mathcal{U}$, if $b \equiv_A b'$, then

$$p \vdash \phi(x; b) \iff p \vdash \phi(x; b').$$

Example 2.16. A type $p(x)$ over a set B is said to be *definable* if for every formula $\phi(x; y)$ without parameters, there is some formula $d\phi(y) \in L(B)$ such that $p \vdash \phi(x; b) \iff b \models d\phi(y)$ for all $b \in B$. We say that p is definable over $A \subseteq B$ if the formula $d\phi(y)$ can be taken to have parameters in A. If the type p is definable, then it is definable over some A of size $\leq |T|$.

Let $p \in S_x(\mathcal{U})$ be definable. Then by the previous remark, it is definable over some $A \subset \mathcal{U}$ of size $\leq |T|$. In particular p is A-invariant.

Example 2.17. A type $p(x)$ is said to be *finitely satisfiable* in a set A if for every formula $\phi(x; b) \in p$, there is $a \in A$ such that $\phi(a; b)$ holds. If $p \in S_x(\mathcal{U})$ is finitely satisfiable in A, then it is A-invariant. Indeed if $\phi(x; y)$ is a formula and $b \equiv_A b'$ are two $|y|$-tuples, then for every element a of A, $a \models \phi(x; b) \leftrightarrow \phi(x; b')$. Hence also $p \vdash \phi(x; b) \leftrightarrow \phi(x; b')$.

We present two constructions to obtain such types.

(1) Let $A \subset \mathcal{U}$ be any small set and let $\mathcal{D}$ be an ultrafilter on $A^{|x|}$. We define $p_{\mathcal{D}} \in S_x(\mathcal{U})$ by:

$$p_{\mathcal{D}} \vdash \phi(x; b) \iff \phi(A; b) \in \mathcal{D},$$

for every formula $\phi(x; b) \in L(\mathcal{U})$. Then $p_{\mathcal{D}}$ is finitely satisfiable in A. Conversely, every global type finitely satisfiable in A is equal to $p_{\mathcal{D}}$ for some (not necessarily unique) ultrafilter $\mathcal{D}$. In particular note that if we take $\mathcal{D}$ to be a principal ultrafilter, then we obtain a realized type.

(2) Assume that T is NIP and let $I = (a_i : i \in \mathcal{I})$ be an indiscernible sequence. Then by Proposition 2.8, the sequence $(\text{tp}(a_i/\mathcal{U}) : i \in \mathcal{I})$ converges in $S(\mathcal{U})$ to some type, called the *limit type* of the sequence I and denoted by $\lim(I)$. This type is finitely satisfiable in I, and indeed in any cofinal subsequence of I.

If $p_0 \in S_x(M)$ is a type, then a *coheir* of p_0 is a global extension of p which is finitely satisfiable in M. Such a coheir always exists: extend $\{\phi(M; b) : \phi(x; b) \in p_0\}$ into an ultrafilter $\mathcal{D}$ on $M^{|x|}$ and consider $p_{\mathcal{D}}$. In fact, the same proof shows that any type p finitely satisfiable in some set A extends to a global type finitely satisfiable in A.

Let p be a global A-invariant type. Then to every formula $\phi(x; y) \in L$, y any finite tuples of variables, we can associate the set $D_p\phi \subseteq S_y(A)$ of types q such that $p \vdash \phi(x; b)$ for some (any) $b \models q$. The family $(D_p\phi : \phi \in L)$ is called the (infinitary) *defining schema* of p. It completely determines the A-invariant type p. Notice that the defining schema is an object of small cardinality; in particular, the monster model $\mathcal{U}$ does not appear in it in any way. Indeed given any bigger set $V \supseteq \mathcal{U}$ of parameters, we can define the extension $p|V$ of p over V by setting $p \vdash \phi(x; b)$ if and only if $\text{tp}(b/A) \in D_p\phi$. The reader should check that this construction depends only on p and not on the choice of A.

This procedure can also work if we start with a type over a small model, as long as enough saturation is present: Let $A \subset M$, and assume that for all n, any type in $S_n(A)$ is realized in M. Let $p \in S_x(M)$ be A-invariant in the following sense: if $b \equiv_A b'$ are in M and $\phi(x; y) \in L$, then $p \vdash \phi(x; b) \iff p \vdash \phi(x; b')$. Then we can associate to p a defining schema $(D_p\phi : \phi(x; y) \in L)$ of p as above. In turn this schema defines a global type $\tilde{p} = \{\phi(x; b) : \phi \in L, \text{tp}(b/A) \in D_p\phi\}$. The fact that we have taken M to realize all finitary types over A ensures consistency of $\tilde{p}$. We see that $\tilde{p}$ is the unique A-invariant global extension of p.

In general if $p \in S(M)$ is A-invariant in the sense of the previous paragraph, but M does not realize all types over A (for example if $M = A$), then p may have more than one global A-invariant extension.

LEMMA 2.18. *Let $p \in S(\mathcal{U})$ be an A-invariant type*:

1. *if p is definable, then it is A-definable*;
2. *if p is finitely satisfiable in some small set, then it is finitely satisfiable in any model $M \supseteq A$.*

PROOF. Assume that p is definable. Let $\phi(x; y)$ be any formula and let $d\phi(y) \in L(\mathcal{U})$ be such that for any $b \in \mathcal{U}$, $p \vdash \phi(x; b) \iff b \models d\phi(y)$. As p is A-invariant, the definable set $d\phi(y)$ is invariant under all automorphisms that fix A pointwise. It follows that it is definable over A and therefore so is p.

Assume now that p is finitely satisfiable in some small model N. Let M be any small model containing A. Let $\phi(x; b)$ be any formula in p and let N_1 realize a coheir of $\text{tp}(N/M)$ over Mb. Then by invariance, p is finitely satisfiable in N_1 and in particular $\phi(N_1; b)$ is non-empty. By the coheir hypothesis, $\phi(M; b)$ is non-empty. Therefore p is finitely satisfiable in M. ⊣

2.2.1. Products and Morley sequences. Let $p(x), q(y) \in S(\mathcal{U})$ be two A-invariant types. We define the type $p(x) \otimes q(y) \in S_{xy}(\mathcal{U})$ as $\text{tp}(a, b/\mathcal{U})$ where $b \models q$ and $a \models p|\mathcal{U}b$.

We use here the canonical extension $p|\mathcal{U}b$ of p to $\mathcal{U}b$ as defined before Lemma 2.18.

If one wants to avoid realizing types over the monster, one can also give the following equivalent definition: Given a formula $\phi(x; y) \in L(B)$, $A \subseteq B \subset \mathcal{U}$, we set $p(x) \otimes q(y) \vdash \phi(x; y)$ if $p \vdash \phi(x; b)$ for some (any) $b \in \mathcal{U}$ with $b \models q|_B$.

The following facts are easy to check from the definitions:

FACT 2.19. *If p and q are both A-invariant, then so is the product $p(x) \otimes q(y)$.*

FACT 2.20. *The relation $\otimes$ is associative: $p(x) \otimes (q(y) \otimes r(z)) = (p(x) \otimes q(y)) \otimes r(z)$. Indeed, both products are equal to $\text{tp}(a, b, c/\mathcal{U})$ where $c \models r$, $b \models q|\mathcal{U}c$ and $a \models p|\mathcal{U}bc$.*

However, $\otimes$ need not be commutative:

Example 2.21. Let T be DLO, and take $p = q$ to be the type at $+\infty$. Then $p(x) \otimes q(y) \vdash x > y$ whereas $q(y) \otimes p(x) \vdash x < y$.

Exercise 2.22. Check that in DLO, any two distinct invariant 1-types p and q commute: $p(x) \otimes q(y) = q(y) \otimes p(x)$.

This is no longer true in RCF: If p and q are two invariant 1-types which concentrate on definable cuts (either $\pm\infty$ or $a^{\pm}$), then they do not commute.

Note that in the definition of $p(x) \otimes q(y)$, we did not actually use the fact that q is invariant. Hence more generally, if $p(x)$ is A-invariant and $q(y) \in S(\mathcal{U})$ is any type, then we can define $p(x) \otimes q(y) \in S_{xy}(\mathcal{U})$ as $\text{tp}(a, b/\mathcal{U})$ where $b \models q$ and $a \models p|\mathcal{U}b$. The type $p(x) \otimes q(y)$ defined this way is invariant if and only if q is.

LEMMA 2.23. *If $p(x)$ is a global definable type and $q(y)$ is a global type finitely satisfiable in some small model, then $p(x) \otimes q(y) = q(y) \otimes p(x)$.*

PROOF. Let $M \prec \mathcal{U}$ be a model such that p is definable over M and q is finitely satisfiable in M. Let $\phi(x, y; d) \in L(\mathcal{U})$. By definability of p, there is a formula $d\phi(y; z) \in L(M)$ such that for every $b, d' \in \mathcal{U}$, we have $p \vdash \phi(x, b; d') \iff \models d\phi(b; d')$. Let $\varepsilon \in \{0, 1\}$ such that $q \vdash d\phi(y; d)^{\varepsilon}$.

Assume for example that $q(y) \otimes p(x) \vdash \phi(x, y; d)$. Let $a \models p$. Then $q(y)|\mathcal{U}a \vdash \phi(a, y; d)$ and as q is finitely satisfiable in M, there is $b \in M$ such that $b \models \phi(a, y; d) \wedge d\phi(y; d)^{\varepsilon}$. Then $p \vdash \phi(x, b; d) \wedge d\phi(b; d)^{\varepsilon}$ and by definition of $d\phi$, this implies that $\varepsilon = 1$.

Therefore $q \vdash d\phi(y;d)$ and it follows that $p(x) \otimes q(y) \vdash \phi(x, y; d)$. $\dashv$

If $p(x)$ is an A-invariant type, we define by induction on $n \in \mathbb{N}^*$:

$$p^{(1)}(x_0) = p(x_0) \quad \text{and} \quad p^{(n+1)}(x_0, \dots, x_n) = p(x_n) \otimes p^{(n)}(x_0, \dots, x_{n-1}).$$

Let also $p^{(\omega)}(x_0, x_1, \dots) = \bigcup p^{(n)}$. For any $B \supseteq A$, a realization $(a_i : i < \omega)$ of $p^{(\omega)}|_B$ is called a *Morley sequence* of p over B (indexed by ω). It follows from associativity of $\otimes$ that such a sequence $(a_i : i < \omega)$ is indiscernible over B (indeed for any $i_1 < \cdots < i_n \in \omega$, we have $\text{tp}(a_{i_1}, \dots, a_{i_n}/B) = p^{(n)}|_B$).

More generally, any sequence indiscernible over B whose EM-type over B is given by $\{p^{(n)}|_B : 1 \leq n < \omega\}$ is called a Morley sequence of p over B.

Exercise 2.24. Let p, q be global A-definable types (resp. global types finitely satisfiable in A), then $p(x) \otimes q(y)$ is also A-definable (resp. finitely satisfiable in A). In particular, so are $p^{(n)}$ and $p^{(\omega)}$.

Exercise 2.25. (T is NIP.) Let $I = (a_t : t \in \mathcal{I})$ be an endless indiscernible sequence. Recall the definition of the limit type $p = \lim(I)$ from Example 2.17. Let $\mathcal{J}$ be any linear order and $J = (b_t : t \in \mathcal{J})$ be a sequence such that $b_t \models p \restriction I\bar{b}_{>t}$. So if we reverse the order of the sequence J, we obtain a Morley sequence of p over I. Show that $I + J$ is indiscernible.

An application to definable groups.

LEMMA 2.26. *(T is NIP.) Let G be a definable group. Let p, q be invariant types concentrating on G such that both $p(x) \otimes q(y)$ and $q(y) \otimes p(x)$ imply $x \cdot y = y \cdot x$. Then $a \cdot b = b \cdot a$ for any $a \models p$ and $b \models q$.*

PROOF. By compactness, there is a small model M over which p and q are invariant and such that $a \cdot b = b \cdot a$ for any (a, b) realizing one of $p \otimes q \restriction M$ or $q \otimes p \restriction M$

We try to build by induction a sequence $(a_n \cdot b_n : n < \omega)$ such that $a_n \models p \restriction Ma_{<n}b_{<n}$, $b_n \models q \restriction Ma_{<n}b_{<n}$ and $a_n \cdot b_n \neq b_n \cdot a_n$. Assume that we succeed. Then by hypothesis, $a_n \cdot b_m = b_m \cdot a_n$ for $n \neq m$. For any $I \subset \omega$ finite, define $b_I = \prod_{n \in I} b_n$. We have that $a_n \cdot b_I = b_I \cdot a_n$ if and only if $n \notin I$. This contradicts NIP.

Therefore the construction must stop and for some n, we cannot find a_n, b_n as required. Then if we let $p_0 = p \restriction Ma_{<n}b_{<n}$ and $q_0 = q \restriction Ma_{<n}b_{<n}$, we have $p_0(x) \wedge q_0(y) \to (x \cdot y = y \cdot x)$. $\dashv$

PROPOSITION 2.27. *(T is NIP.) Let G be a definable group, and assume that there is a small subset $A \subset G$ such that any two elements of A commute, then there is a definable abelian subgroup of G containing A.*

PROOF. Let $S_A \subset S(\mathcal{U})$ be the subset of global 1-types finitely satisfiable in A. This is a closed set of the space of types, and as such it is compact. For any $p, q \in S_A$, the pair (p, q) satisfies the hypothesis of the previous lemma. Hence by compactness, there are formulas $\phi_{p,q}(x)$ and $\psi_{p,q}(y)$ such that $p(x) \vdash \phi_{p,q}(x)$, $q \vdash \psi_{p,q}(y)$ and $\phi_{p,q}(x) \wedge \psi_{p,q}(y) \to (x \cdot y = y \cdot x)$.

For a given p, the family $(\psi_{p,q}(y) : q \in S_A)$ of clopens covers S_A. Hence there is a finite subfamily $\psi_{p,q_1}(x), \dots, \psi_{p,q_n}(x)$ which already covers it. Set $\phi_p(x) = \bigwedge_{i \leq n} \phi_{p,q_n}(x)$ and $\psi_p(y) = \bigvee_{i \leq n} \psi_{p,q_n}(y)$. Then again, $p(x) \vdash \phi_p(x)$ and we can find a finite family $\phi_{p_1}(x), \dots, \phi_{p_k}(x)$ which cover S_A. Set $\phi(x) = \bigvee_{i \leq k} \phi_{p_i}(x)$ and $\psi(y) = \bigwedge_{i \leq k} \psi_{p_i}(y)$.

Then $\phi(x) \wedge \psi(y) \to (x \cdot y = y \cdot x)$ and all types of S_A concentrate on both $\phi(x)$ and $\psi(y)$. In particular this is true for realized types $x = a$, $a \in A$. Let H be the subgroup of G defined by $C_G(C_G(\phi \wedge \psi))$, where $C_G(X) = \{a \in G : a \cdot x = x \cdot a \text{ for all } x \in X\}$. Then H is a definable abelian subgroup of G containing A. ⊣

Note that if A is finite, then the proposition is easy and does not require NIP: take $H = C_G(C_G(A))$.

2.2.2. Generically stable types. In this section and the following one we assume that T is NIP (except in Proposition 2.43 where we give yet another characterization of NIP).

LEMMA 2.28. *(T is NIP.) Let $(a_i : i \in \mathfrak{I})$ be a totally indiscernible sequence of $|x|$-tuples, and $\phi(x; b) \in L(\mathcal{U})$ a formula, then the set $\mathfrak{I}_\phi = \{i \in \mathfrak{I} : \models \phi(a_i; b)\}$ is finite or cofinite in $\mathfrak{I}$.*

PROOF. Otherwise, we can build a sequence $(i_k : k < \omega)$ of pairwise distinct members of $\mathfrak{I}$ such that $i_k \in \mathfrak{I}_\phi$ if and only if k is even. Then the sequence $(a_{i_k} : k < \omega)$ is indiscernible and the formula $\phi(a_{i_k}; b)$ holds if and only if k is even. This contradicts Proposition 2.8. ⊣

Note that it follows from the proof that the cardinality of either $\{i \in \mathfrak{I} : \models \phi(a_i; b)\}$ or its complement is bounded by $\mathrm{alt}(\phi(x; y))/2$.

THEOREM 2.29. *(T is NIP.) Let p be a global A-invariant type. Then the following are equivalent:*

(i) *$p = \lim(I)$ for any $I \models p^{(\omega)}|_A$;*
(ii) *p is definable and finitely satisfiable in some small M;*
(iii) *$p_x \otimes p_y = p_y \otimes p_x$;*
(iv) *any Morley sequence of p is totally indiscernible.*

PROOF. (i) ⇒ (ii): Assume (i). Then p is finitely satisfiable in I. Also, let $\phi(x; y) \in L$ and let $b \in \mathcal{U}$. Then $p \vdash \phi(x; b)$ if and only if there is some sequence $(a_k : k < \omega) \models p^{(\omega)}|_A$ such that $\phi(a_k; b)$ holds for all k. Therefore $\{b \in \mathcal{U} : p \vdash \phi(x; b)\}$ is a type-definable set. The same holds for $\neg\phi$ instead of ϕ, so the complement of that set is also type-definable. It follows that it is definable. Therefore p is a definable type.

(ii) ⇒ (iii): Follows from Lemma 2.23.

(iii) ⇒ (iv): By associativity of $\otimes$, and writing any permutation as a product of transpositions, we see that for every n and $\sigma \in Sym(n)$, $p(x_{n-1}) \otimes \cdots \otimes p(x_0) =$

$p(x_{\sigma(n-1)}) \otimes \cdots \otimes p(x_{\sigma(0)})$. It follows that the Morley sequence of p is totally indiscernible.

(iv) $\Rightarrow$ (i): Let $\phi(x;b) \in p$ and $I \models p^{(\omega)}|_A$. Let $J \models p^{(\omega)}|_{AIb}$. Then $I + J$ is a totally indiscernible sequence. Then all points of J satisfy $\phi(x;b)$. Therefore by Lemma 2.28, at most a finite number of points of I satisfy $\neg\phi(x;b)$. Hence $\lim(I) \vdash \phi(x;b)$ as required. ⊣

DEFINITION 2.30. (T is NIP.) An invariant type satisfying the equivalent conditions of Theorem 2.29 is called *generically stable*.

Example 2.31.

1. Consider the language $L = \{R_n(x, y) : n < \omega\}$ and the L-structure M whose universe is $\mathbb{Q}$ and such that $M \models R_n(x, y)$ if and only if $x < y$ and $|x - y| < n$. This is sometimes called a "local order".

   Let $p(x)$ be the global $\emptyset$-invariant type satisfying $\neg R_n(a, x) \wedge \neg R_n(x, a)$ for all $a \in \mathcal{U}$. Then this type is generically stable.
2. Consider the two-sorted theory of vector spaces over a real-closed field: We have one sort R for the field equipped with the usual ordered field structure and one sort V for the vector space equipped with the group structure. Finally, there is an additional binary function symbol from $R \times V$ to V for scalar multiplication. The axioms say that R is real closed and that V is an infinite dimensional R-vector space. Let p be the generic type of V, namely the type which does not lie in any $\mathcal{U}$-definable proper sub-vector space. Then p is $\emptyset$-invariant and generically stable. Apart from realized types, it is the only generically stable type in this structure.
3. In the theory T_{dt} of dense meet-trees, the generic type of a closed cone is generically stable (see Section 2.3.1).
4. In the theory ACVF of algebraically closed valued fields, the generic type of a closed ball is generically stable.

Remark 2.32. If p is generically stable, then it is definable and finitely satisfiable over any one of its Morley sequences. (Because if I is a Morley sequence of p, then $p = \lim(I)$ is finitely satisfiable in I, furthermore it is I-invariant, so it is definable over I). We can write a definition explicitly. Let $\phi(x; y)$ be a formula, and let $N = \mathrm{alt}(\phi)$. If $(a_k : k < \omega)$ is a Morley sequence of p, and $b \in \mathcal{U}$, then

$$p \vdash \phi(x;b) \iff \bigvee_{\substack{A \subset N+1 \\ |A| > N/2}} \bigwedge_{k \in A} \phi(a_k; b).$$

Notice in particular that the form of the formula giving the ϕ-definition depends only on ϕ and not on the type p.

PROPOSITION 2.33. (T *is NIP.*) *Let p be generically stable and q any invariant type. Then $p(x) \otimes q(y) = q(y) \otimes p(x)$.*

PROOF. Assume for a contradiction that for some formula $\phi(x;y) \in L(\mathcal{U})$ we have $p(x) \otimes q(y) \vdash \phi(x;y)$ and $q(y) \otimes p(x) \vdash \neg\phi(x;y)$. Let $(a_k : k < \omega) \models p^{(\omega)}$, $b \models q|\mathcal{U}a_{<\omega}$ and $(a_k : \omega \leq k < \omega 2) \models p^{(\omega)}|\mathcal{U}a_{<\omega}b$. Then for $k < \omega$, $\neg\phi(a_k;b)$ holds and for $k \geq \omega$, we have $\phi(a_k;b)$. As the sequence $(a_k : k < \omega 2)$ is totally indiscernible, this contradicts Lemma 2.28. ⊣

LEMMA 2.34. *(T is NIP.) Let p be generically stable, then for any integer n, $p^{(n)}$ is also generically stable.*

PROOF. This follows for example from Exercise 2.24. ⊣

PROPOSITION 2.35. *(T is NIP.) Let p be a generically stable, A-invariant type. Then p is the unique A-invariant extension of $p|_A$.*

PROOF. Notice that if q is any A-invariant type and $r, s \in S(\mathcal{U})$ are such that $r|_A = s|_A$, then $(q \otimes r)|_A = (q \otimes s)|_A$.

Let q be any A-invariant extension of $p|_A$. First we show by induction that $q^{(n)}|_A = p^{(n)}|_A$. The case $n = 1$ is the hypothesis on q. Assume that $q^{(n)}|_A = p^{(n)}|_A$. We have, using Proposition 2.33 on line 3:

$$\begin{aligned} q^{(n+1)}(x_1,\dots,x_{n+1})|_A &= (q(x_{n+1}) \otimes q^{(n)}(x_1,\dots,x_n))|_A \\ &= (q(x_{n+1}) \otimes p^{(n)}(x_1,\dots,x_n))|_A \\ &= (p^{(n)}(x_1,\dots,x_n) \otimes q(x_{n+1}))|_A \\ &= (p^{(n)}(x_1,\dots,x_n) \otimes p(x_{n+1}))|_A \\ &= p^{(n+1)}(x_1,\dots,x_{n+1})|_A. \end{aligned}$$

Hence $q^{(\omega)}|_A = p^{(\omega)}|_A$ and therefore any Morley sequence of q is totally indiscernible. It follows that q is generically stable and that $q = \lim(I)$ for any $I \models q^{(\omega)}|_A$. But also $p = \lim(I)$ for any $I \models p^{(\omega)}|_A = q^{(\omega)}|_A$. Hence $p = q$. ⊣

2.2.3. Eventual types. We have seen that any generically stable type is the average of any one of its Morley sequences. This is no longer true for general invariant types. We show now that we can nevertheless recover an A-invariant type from its Morley sequence over A although in a slightly more complicated way.

PROPOSITION 2.36. *(T is NIP.) Let $p, q \in S_x(\mathcal{U})$ be A-invariant types. If $p^{(\omega)}|_A = q^{(\omega)}|_A$, then $p = q$.*

PROOF. By assumption, if I is a Morley sequence of p over A (indexed by any linear order) then it is also a Morley sequence of q over A. Let $\phi(x;d) \in L(\mathcal{U})$ be any formula and let $I = (a_i : i \in \mathfrak{I})$ be a Morley sequence of p over A such that $\mathrm{alt}(\phi(x;d), I)$ is maximal. Let $\varepsilon \in \{0,1\}$ be such that $\lim(I) \vdash \phi(x;d)^\varepsilon$. Now let $a \models p|_{AId}$. Then the sequence $I + (a)$ is also a Morley sequence of p over A. By maximality of I, we must have $\mathrm{alt}(\phi(x;d), I) = \mathrm{alt}(\phi(x;d), I + (a))$. This implies that $a \models \phi(x;d)^\varepsilon$. Therefore $p \vdash \phi(x;d)^\varepsilon$ and as the roles of p and q are interchangeable, we also have $q \vdash \phi(x;d)^\varepsilon$. Thus $p = q$. ⊣

DEFINITION 2.37. Let $A \subset \mathcal{U}$ and let I be an indiscernible sequence. We say that I is *based* on A if I is indiscernible over A and if for any two A-indiscernible sequences I_1, I_2 of same EM-type as I over A, there is some tuple a such that $I_1 + (a)$ and $I_2 + (a)$ are both indiscernible over A.

PROPOSITION 2.38. *(T is NIP.) Let I be any indiscernible sequence based on A. There is a unique global type p with the following property: For any sequence $J \equiv^{EM}_A I$ and any $B \subset \mathcal{U}$, there is $a \in \mathcal{U}$ such that $J + (a)$ is A-indiscernible and $a \models p|_B$. Moreover, p is invariant over A.*

PROOF. We first show uniqueness. Assume that there were two such types p and q and pick a formula $\phi(x; b) \in L(\mathcal{U})$ such that $p \vdash \phi(x; b)$ and $q \vdash \neg\phi(x; b)$. Now choose by induction tuples $(a_i : i < \omega)$ such that:

- $I + (a_i : i < \omega)$ is A-indiscernible;
- for i even, $a_i \models p|_{Ab}$;
- for i odd, $a_i \models q|_{Ab}$.

This is possible by hypothesis. In the end, the formula $\phi(x; b)$ alternates infinitely often on the sequence $I + (a_i : i < \omega)$ contradicting NIP.

We now show existence. Let $\phi(x; d) \in L(\mathcal{U})$ be any formula. Let $I_1 \equiv^{EM}_A I$ and $I_2 \equiv^{EM}_A I$ such that $\mathrm{alt}(\phi(x; d), I_i)$ is maximal, for $i \in \{1, 2\}$. Let ε_i, for $i = 1, 2$ be such that $\lim(I_i) \vdash \phi(x; d)^{\varepsilon_i}$. As I is based on A, there is some tuple a such that both $I_1 + (a)$ and $I_2 + (a)$ are indiscernible over A. By maximality of I_1, we have $\mathrm{alt}(\phi(x; d), I_1 + (a)) = \mathrm{alt}(\phi(x; d), I_1)$ and $a \models \phi(x; d)^{\varepsilon_1}$. Similarly we have $a \models \phi(x; d)^{\varepsilon_2}$ therefore $\varepsilon_1 = \varepsilon_2$.

We now define a global type p as follows: pick some $J \equiv^{EM}_A I$ such that $\mathrm{alt}(\phi(x; d), J)$ is maximal, and set $\mathrm{Ev}(I/A) \vdash \phi(x; d)$ if and only if $\lim(J) \vdash \phi(x; d)$. By the previous argument, this does not depend on the choice of J. It is clear that this defines a consistent, complete type over $\mathcal{U}$. Also, p only depends on the type of I over A (indeed of its EM-type) therefore p is an A-invariant type. ⊣

DEFINITION 2.39. The type p defined above is called the *eventual type* of I over A. We will denote it by $\mathrm{Ev}(I/A)$.

Example 2.40. Let T be DLO, and consider an increasing sequence $I = (a_i : i < \omega)$. Let $b_2 < b_1$ be two points greater than all the a_i's. Set $A_1 = \{b_1\}$ and $A_2 = \{b_1, b_2\}$, then I is indiscernible both over A_1 and A_2. The type $\mathrm{Ev}(I/A_1)$ is the definable type "b_1^-" whereas $\mathrm{Ev}(I/A_2)$ is the type "b_2^-".

Now take A to be a set lying above all the a_i's with no smallest element. Then $\mathrm{Ev}(I/A)$ is axiomatized by $\{x < a : a \in A\} \cup \{x > b : b < A\}$. It is not definable, but it is finitely satisfiable in any model containing A (or indeed a coinitial segment of A).

We see from that example that the properties of the type $\mathrm{Ev}(I/A)$ really depend on both I and A and not only on I. However the EM-type (over $\emptyset$) of a Morley sequence of $\mathrm{Ev}(I/A)$ does not depend on A:

LEMMA 2.41. *(T is NIP.) If I is an indiscernible sequence based on A, then I is a Morley sequence of* $\mathrm{Ev}(I/A)$ *over A.*

PROOF. Without loss, we may assume that $I = (a_i : i < \omega)$. It is enough to show that for each $n < \omega$, we have $a_n \models \mathrm{Ev}(I/A) \upharpoonright Aa_{<n}$. By definition of the eventual type, we can find a_* such that $J = I + (a_*)$ is A-indiscernible and $a_* \models \mathrm{Ev}(I/A) \upharpoonright AI$. By indiscernibility of J, $\mathrm{tp}(a_n/Aa_{<n}) = \mathrm{tp}(a_*/Aa_{<n}) = \mathrm{Ev}(I/A) \upharpoonright Aa_{<n}$. ⊣

PROPOSITION 2.42. *(T is NIP.) There is a bijection between EM-classes over A of indiscernible sequences based on A and A-invariant types.*

PROOF. From left to right, we map an indiscernible sequence I to the type $\mathrm{Ev}(I/A)$ which only depends on the EM-type of I over A. In the other direction, we map an A-invariant type p to the EM-class of its Morley sequence. ⊣

PROPOSITION 2.43. *The theory T is NIP if and only if for all $M \models T$, for all $p \in S(M)$, p has at most $2^{|M|+|T|}$ global coheirs.*

PROOF. We show left to right. Assume that T is NIP and let $M \models T$, then by Proposition 2.36 an M-invariant type is determined by $p^{(\omega)}|_M$. There are at most $2^{|M|+|T|}$ values for that type, so there are at most $2^{|M|+|T|}$ global M-invariant types. The result follows.

Conversely, assume that T has IP and set $\lambda = |T|$. Then there is a set $\{a_i : i < \lambda\}$ of finite tuples and a formula $\phi(x; y)$ such that for any $A \subseteq \lambda$, we can find some b_A such that: $\phi(a_i; b_A) \iff i \in A$. Let M be a model of size λ containing all the a_i's. For $\mathcal{D}$ an ultrafilter over λ, we define $p_{\mathcal{D}}$ as the average of $\mathrm{tp}(a_i/\mathcal{U})$ along $\mathcal{D}$. Then we have $p_{\mathcal{D}} \vdash \phi(x; b_A) \iff A \in \mathcal{D}$. This shows that for $\mathcal{D} \neq \mathcal{D}'$, the two types $p_{\mathcal{D}}$ and $p_{\mathcal{D}'}$ are distinct. Furthermore, $p_{\mathcal{D}}$ is finitely satisfiable in M. As there are 2^{2^λ} pairwise distinct ultrafilters on λ, we have obtained 2^{2^λ} global types finitely satisfiable in M. Since there are only 2^λ types in variable x over M, there is at least one such type which has 2^{2^λ} global coheirs. ⊣

Example 2.44. Let M be a model of DLO and let $p \in S_1(M)$ be a type corresponding to a cut of M of infinite cofinalities from both sides (equivalently, p is a non-definable type over M). Then p has exactly two global coheirs: one sticking to the left of the cut and the other one sticking to the right of the cut. In fact those are the only global M-invariant extensions of p.

Exercise 2.45. Let $(a_i : i < \omega 2)$ be indiscernible. Show that $(a_i : i < \omega)$ is based on $(a_i : \omega \leq i < \omega 2)$. Conclude that any indiscernible sequence is based on some set.

Exercise 2.46. Let $I = (a_i : i < \omega)$ be a totally indiscernible sequence. Define $Cb(I)$ to be the intersection $\bigcap_F \mathrm{acl}^{eq}(\{a_i : i \in F\})$ where F ranges over all infinite subsets of ω.

1. Show that $\lim(I)$ is definable over $Cb(I)$ and that any automorphism of $\mathcal{U}$ fixes the type $\lim(I)$ if and only if it fixes $Cb(I)$ pointwise.
2. Show that I is based on a set A if and only if it is indiscernible over A and $Cb(I) \subseteq \mathrm{dcl}^{eq}(A)$.

2.3. Additional topics

2.3.1. Case study: dense trees. Trees—or structures built around trees—are a major source of examples of NIP theories. In this section, we concentrate on the theory of dense meet-trees which is the model companion of finite meet-trees.

DEFINITION 2.47 (Tree). A *tree* is a partially ordered set $(M, \leq)$ such that for every $a \in M$, the set $\{x \in M : x \leq a\}$ is linearly ordered by $\leq$ and for any $a, b \in M$, there is some c smaller or equal to both a and b.

We say that $(M, \leq)$ is a *meet-tree* if in addition: for every two points $a, b \in M$, the set $\{x \in M : x \leq a \wedge x \leq b\}$ has a greatest element, which we denote by $a \wedge b$. (In other words, $(M, \leq)$ is a meet-semilattice.)

A *leaf* of the tree M is a point in M which is maximal.

Note in particular that with this definition, a linearly ordered set is a tree.

Exercise 2.48. Let $(M, \leq)$ be a tree. Then there is a meet-tree $(\hat{M}, \leq)$ which has M as a subtree, such that for every $a < b \in \hat{M}$, there is $c \in M$, $a \leq c \leq b$. Furthermore, $\hat{M}$ is interpretable (along with the embedding of M into it) in the structure $(M, \leq)$.

Exercise 2.49. If $(M, \leq)$ is a meet-tree, then we have, for any $a, b, c \in M$:

$$(a \wedge b) \wedge c = a \wedge (b \wedge c) = \min(a \wedge c, b \wedge c).$$

Let $(M, \leq)$ be a meet-tree and $c \in M$ a point. The *closed cone* of center c is by definition the set $C(c) := \{x \in M : x \geq c\}$. We can define on $C(c)$ a relation E_c by: xE_cy if $x \wedge y > c$. It follows from Exercise 2.49 that this is an equivalence relation. We define an *open cone* of center c to be a equivalence class under the relation E_c.

The theory of meet-trees in the language $\{\leq, \wedge\}$ has a model-companion, namely the theory T_{dt} of dense meet-trees which is defined by the following axioms:

- $\leq$ defines a meet-tree on the universe, and $\wedge$ is the meet relation;
- for any point c, $\{x : x \leq c\}$ is dense with no first element;
- for any point c, there are infinitely many open cones of center c.

Exercise 2.50. The theory T_{dt} is complete and admits elimination of quantifiers in the language $\{\leq, \wedge\}$.

In fact, T_{dt} is $\aleph_0$-categorical and the unique countable model is the Fraïssé limit of finite meet-trees.

PROPOSITION 2.51. *The theory T_{dt} is NIP.*

PROOF. (Sketch) By quantifier elimination and Lemma 2.9, we only need to check that the formulas $x \leq y$, $x \wedge y_1 = y_2$, $y_1 \wedge y_2 = x$ are NIP, where the names of the variables indicate the intended separation into variables and parameters. We leave the verifications to the reader. $\dashv$

We want to understand invariant 1-types in T_{dt}. We first study indiscernible sequences of points. So let $I = (a_i : i < \omega)$ be an indiscernible sequence of points. By inspection, we see that there are exactly 6 possibilities for the EM-type of I over $\emptyset$ (the reader is strongly advised to make drawings of the different configurations):

(0) The sequence I is constant: $a_i = a_j$, $i, j < \omega$.
(Ia) The sequence I is increasing: $a_i < a_j$, $i < j < \omega$.
(Ib) The sequence I is decreasing: $a_j < a_i$, $i < j < \omega$.
(II) Fan: Elements of I are pairwise incomparable and there is some $c \in \mathcal{U}$ such that $a_i \wedge a_j = c$ for all $i, j < \omega$.
(IIIa) Increasing comb: Elements of I are pairwise incomparable, $a_i \wedge a_j$, $i < j$, depends only on i and is increasing with i.
(IIIb) Decreasing comb: Elements of I are pairwise incomparable, $a_i \wedge a_j$, $i < j$, depends only on j and is decreasing with j.

We now study what types can appear as eventual types of those sequences. We treat two cases and leave the others as exercise.

(Ia) Let $I = (a_i : i < \omega)$ be an increasing sequence and A any set of parameters over which I is indiscernible. If A contains a point b greater than all the a_i's, then I is based on A: Given some $J \equiv_A^{EM} I$, both I and J lie below b. As the points below b are linearly ordered, we can find some a greater than all elements of both I and J and smaller than all elements of A lying below b (a set which we will denote by $A_{\leq b}$). Then $I + (a)$ and $J + (a)$ are both A-indiscernible. When this is the case, the eventual type $\mathrm{Ev}(I/A)$ is axiomatized by $\{x < d : d \in A_{\leq b}\} \cup \{x > d : d \leq b, d < A_{\leq b}\}$. We know that $\mathrm{Ev}(I/A)$ cannot be both definable and finitely satisfiable in some small model since its Morley sequence is not totally indiscernible. More precisely, if $A_{\leq b}$ has a smallest element c, then $\mathrm{Ev}(I/A)$ is definable over c and is not finitely satisfiable in any small model. If $A_{\leq b}$ does not have a smallest element, then $\mathrm{Ev}(I/A)$ is not definable, but is finitely satisfiable in any model containing A (this is the same situation as for an increasing sequence in DLO).

If A does not contain a point greater than all the a_i's, then I is not based on A: Take a sequence $J = (b_i : i < \omega)$ with $b_0 = a_0$, but b_1 and a_1 are incomparable. The hypothesis on A implies that J is indiscernible over it and has same EM-type as I. However, there is no $a \in \mathcal{U}$ which is above both a_1 and b_1.

(II) If I is a fan centered at c, then it is a totally indiscernible sequence. Its limit type $\lim(I)$ is generically stable: it is definable over c and finitely satisfiable in any model containing c. We call it the *generic of the closed cone centered at c*. The sequence I is based on a set A if and only if it is indiscernible over A and $c \in \operatorname{dcl}(A)$ (otherwise one can move I over A to a fan centered on some $c' \neq c$).

Exercise 2.52. Carry out a similar analysis for each of the remaining cases: determine over which A the sequence I is based and describe the eventual type $\operatorname{Ev}(I/A)$ whenever it is defined.

In particular, observe that for any indiscernible sequence I in T_{dt}, there is some A such that $\operatorname{Ev}(I/A)$ is a definable type.

2.3.2. Stability.

DEFINITION 2.53. We say that a formula $\phi(x;y)$ has the order property if there are sequences $(a_i : i < \omega)$ and $(b_i : i < \omega)$ such that:

$$\mathcal{U} \models \phi(a_i; b_j) \iff i \leq j.$$

Observe that this property is invariant under changing ϕ to ϕ^{opp}.

Recall that if $\phi(x;y)$ is a formula, then a ϕ-type over A is a maximal consistent set of formulas of the form $\phi(x;a)$ or $\neg\phi(x;a)$ with $a \in A$. A ϕ-type p over A is *definable* if there is a formula $d_p\phi(y;b)$ with $b \in A$ such that for any $a \in A$, we have $p \vdash \phi(x;a) \iff \mathcal{U} \models d_p\phi(a;b)$.

DEFINITION 2.54. We say that a formula $\phi(x;y)$ has the binary tree property if there are $|y|$-tuples $(b_f : f \in {}^{<\omega}2)$ such that for every $f \in {}^{\omega}2$, the partial type

$$\{\phi(x; b_{f\restriction i})^{f(i)} : i < \omega\}$$

is consistent.

PROPOSITION 2.55. *Let $\phi(x;y)$ be a formula. The following are equivalent:*

(i) *for any infinite set $|A|$, $|S_\phi(A)| \leq |A|$;*
(ii) *for some infinite cardinal λ, for every set A of size λ, $|S_\phi(A)| \leq |A|$;*
(iii) *$\phi(x;y)$ does not have the order property;*
(iv) *$\phi(x;y)$ does not have the binary tree property;*
(v) *Any ϕ-type over a set A is definable.*

PROOF. See e.g. [115, 8.2.3, 8.3.1]. ⊣

DEFINITION 2.56. A formula satisfying the equivalent conditions of Proposition 2.55 is called *stable*.

A theory T is stable if all formulas are.

Note that in the definition of a stable theory it does not matter if we quantify over formulas with or without parameters since if the formula $\phi(x; y, d)$ is unstable, then so is the formula $\phi(x; y\hat{}z)$. Also, it is easy to see, using for example the order property, that a stable formula is NIP.

Remark 2.57. It follows from the proof of Proposition 2.55 (iv) that if ϕ is stable, there is a finite set $d_1\phi(y; z), \dots, d_n\phi(y; z)$ such that any ϕ-type over any set A is definable by an instance of one of the $d_i\phi$'s.

The following proposition gives more details on the formula defining a ϕ-type in the case where the type is finitely-satisfiable in the parameter set A. This is always the case for example when A is a model.

PROPOSITION 2.58. *Let $\phi(x; y)$ be stable and let $p \in S_\phi(A)$ be a ϕ-type which is finitely satisfiable in A. Then p is definable by a positive Boolean combination of formulas of the form $\phi(a; y)$ for $a \in A$. Furthermore, the size of this Boolean combination is bounded by an integer depending only on ϕ.*

PROOF. See e.g. [21, Lemma 2.10]. Uniformity is not stated explicitly there, but follows from the proof. ⊣

LEMMA 2.59. *A theory T is stable if and only if all indiscernible sequences are totally indiscernible.*

PROOF. Assume that T is unstable and that $\phi(x; y)$ has the order property as witnessed by $I = (a_i, b_i : i < \omega)$. Let $I' = (a'_i, b'_i : i < \omega)$ be indiscernible and realize the EM-type of I. Then we have $\phi(a'_i, b'_j)$ if and only if $i \leq j$, hence I' is not totally indiscernible.

Conversely, assume that there is some $I = (a_i : i < \omega)$ which is indiscernible, but not totally. Then there is a formula $\phi(x, y)$ over some parameters A such that $\phi(a_i, a_j)$ holds if and only if $i \leq j$. Hence T is not stable. ⊣

THEOREM 2.60. *For any theory T, the following are equivalent:*

(i) *T is stable;*
(ii) *for some cardinal λ, for any model M of size λ, $|S(M)| \leq \lambda$;*
(iii) *for any model M, $|S(M)| \leq |M|^{|T|}$;*
(iv) *every type over any model is definable.*
(v) *every type over any set is definable.*

PROOF. (i) ⇒ (v): Follows from Proposition 2.55.

(v) ⇒ (iv): Clear.

(iv) ⇒ (iii): Given a formula $\phi(x; y)$, there can be at most $|M|$ definitions for a ϕ-type over M. Hence if all types are definable, there are at most $|M|^{|T|}$ many types over M.

(iii) ⇒ (ii): Take λ such that $\lambda^{|T|} = \lambda$.

(ii) ⇒ (i): Assume that the formula $\phi(x; y)$ has the order property as witnessed by $(a_i, b_i : i < \omega)$. Let λ be any cardinal. We can find a linear order $\mathfrak{I}$ of size λ such that its completion $\bar{\mathfrak{I}}$ has size $> \lambda$ (Exercise 2.72). Let

$I = (a'_i, b'_i : i \in \mathfrak{I})$ be an indiscernible sequence realizing the EM-type of $(a_i, b_i : i < \omega)$ and let M be a model of size λ containing I. For any $c \in \bar{\mathfrak{I}}$, the partial type $p_c = \{\phi(x; b'_i) : i < c\} \cup \{\neg\phi(x; b'_j) : j > c\}$ is consistent. Thus we obtain more than λ types over M. ⊣

Example 2.61. Here are some classical examples of stable theories:

- Any theory of equivalence relations $\{E_i : i \in \mathfrak{I}\}$ which eliminates quantifiers in this language is stable.
- If R is a ring, an R-module can be seen as a structure $(M; 0, +, r)_{r \in R}$, where r is a unary function symbol interpreted as scalar multiplication by $r \in R$. In this sense, any R-module is stable. In particular the theory of pure abelian groups is stable (up to bi-definability, it coincides with the theory of $\mathbb{Z}$-modules).
- The theory ACF of algebraically closed fields and the theory SCF_p of separably closed fields of characteristic p are stable.
- The theory DCF of differentially closed fields is stable (see e.g. [115], Section 3.3.7).

DEFINITION 2.62. A partial type $\pi(x)$ is said to be *fully stable* if we cannot find a formula $\phi(x; y)$ and a sequence $(a_i, b_i : i < \omega)$ such that all the a_i's realize $\pi(x)$ and such that $\phi(a_i; b_j)$ holds if and only if $i \leq j$.

Remark 2.63. A partial type $\pi(x)$ is fully stable if and only if, for every formula $\phi(x; y)$, there is some $\psi(x) \in \pi(x)$ such that the formula $\psi(x) \wedge \phi(x; y)$ is stable. This follows from compactness.

We see that the definition of fully stable involves the whole structure, as one can take the b_i's anywhere. In general this is stronger than for example asking that all indiscernible sequences of $\pi(x)$ are totally indiscernible (an internal condition on the type). However, in NIP theories, the two coincide.

PROPOSITION 2.64. (T *is NIP.*) *Let* $\pi(x)$ *be a partial type. The following are equivalent*:

(i) $\pi(x)$ *is fully stable*;
(ii) *Any global extension of* $\pi(x)$ *is generically stable*;
(iii) *Any indiscernible sequence of realizations of* $\pi(x)$ *is totally indiscernible.*

PROOF. (iii) ⇔ (i): It is clear that if $\pi(x)$ is fully stable, then any indiscernible sequence of realizations of $\pi(x)$ is totally indiscernible. Conversely, assume that $\pi(x)$ is not fully stable. Then we can find a formula $\phi(x; y)$ and an indiscernible sequence $(a_i, b_i : i \in \mathbb{Q})$ such that $\phi(a_i, b_j)$ holds if and only if $i \leq j$. Then assuming (iii), the sequence $(a_i : i \in \mathbb{Q})$ is totally indiscernible. However, the set $\{i \in \mathbb{Q} : \models \phi(a_i, b_0)\}$ is finite and cofinite. This contradicts Lemma 2.28.

(i) ⇒ (ii):. If $\pi(x)$ is fully stable, then by Remark 2.63 and Proposition 2.55, any extension of $\pi(x)$ to a complete type over any set is definable. In particular any global extension of $\pi(x)$ is definable (hence invariant). As furthermore any indiscernible sequence of realizations of $\pi(x)$ is totally indiscernible, we conclude that any invariant type extending $\pi(x)$ is generically stable.

(ii) $\Rightarrow$ (iii): Let I be an indiscernible sequence of realizations of $\pi(x)$. Then the limit type $\lim(I)$ is a global invariant type extending $\pi(x)$. It is generically stable if and only if I is totally indiscernible. Hence the result. ⊣

2.3.3. The strict order property.

DEFINITION 2.65. We say that a formula $\phi(x; y)$ has the strict order property (SOP) if there exists a sequence $(b_i : i < \omega)$ of $|y|$-tuples such that for all $i < \omega$,

$$\phi(\mathcal{U}; b_i) \subsetneq \phi(\mathcal{U}; b_{i+1}).$$

Assume that the formula $\psi(x_1, x_2)$ defines a preorder (that is, is reflexive and transitive). We say that ψ has infinite chains if we can find a sequence $(a_i : i < \omega)$ of tuples in $\mathcal{U}$ such that $\psi(a_i, a_{i+1}) \wedge \neg\psi(a_{i+1}, a_i)$ holds for all $i < \omega$.

OBSERVATION 2.66. *For a given theory T the following are equivalent:*

- *some formula has SOP;*
- *there is a formula in T defining a preorder with infinite chains;*
- *there is a formula in T^{eq} defining a partial order with infinite chains.*

PROOF. Assume that the formula $\phi(x; y)$ has SOP as witnessed by a sequence $(b_i : i < \omega)$. Then the formula $\psi(y_1, y_2) = \forall x(\phi(x; y_1) \rightarrow \phi(x; y_2))$ defines a preorder for which the sequence $(b_i : i < \omega)$ forms an infinite chain.

Now assume that the formula $\psi(x_1, x_2)$ defines a preorder. Let E be the equivalence relation given by $x_1 E x_2 \iff \models \psi(x_1, x_2) \wedge \psi(x_2, x_1)$. Then ψ induces a definable partial order on the sort of E-equivalence classes which has infinite chains if ψ does.

Finally, assume that E is a definable equivalence relation and $\psi(u_1, u_2)$ defines a partial order with infinite chains on the sort of E-equivalence classes. Then the formula $\phi(x_1; x_2) = \psi(\hat{x}_1, \hat{x}_2)$ has the strict order property, where $\hat{x}$ denotes the E-class of x. ⊣

THEOREM 2.67. *Assume that T is unstable. Then at least one of the following holds:*

- *there is a formula $\phi(x; y)$ which has IP;*
- *there is a formula $\phi(x; y)$ which has SOP.*

PROOF. Let $\phi(x; y)$ be unstable and NIP. There is some indiscernible sequence $(a_i : i < \omega)$ and a sequence $(b_N : N < \omega)$ such that $\phi(a_i; b_N)$ holds if and only if $i < N$. By NIP, there is some integer n and $\eta : n \to \{0, 1\}$ such that

$$\bigwedge_{i<n} \phi(a_i; y)^{\eta(i)}$$

is inconsistent. Starting with that formula, we change one by one instances of $\neg\phi(a_i;y) \wedge \phi(a_{i+1};y)$ to $\phi(a_i;y) \wedge \neg\phi(a_{i+1};y)$. In the end, we arrive at a formula of the form $\bigwedge_{i<N} \phi(a_i;y) \wedge \bigwedge_{N\leq i<n} \neg\phi(a_i;y)$. The tuple b_N satisfies that formula. There is therefore one step in the process in which we pass from an inconsistent formula to a consistent one. Namely, there is some $i_0 < n$, $\eta_0 : n \to \{0,1\}$ such that

$$\bigwedge_{i<i_0} \phi(a_i;y)^{\eta(i)} \wedge \neg\phi(a_{i_0};y) \wedge \phi(a_{i_0+1};y) \wedge \bigwedge_{i_0+1<i<n} \phi(a_i;y)^{\eta(i)}$$

is inconsistent, but

$$\bigwedge_{i<i_0} \phi(a_i;y)^{\eta(i)} \wedge \phi(a_{i_0};y) \wedge \neg\phi(a_{i_0+1};y) \wedge \bigwedge_{i_0+1<i<n} \phi(a_i;y)^{\eta(i)}$$

is consistent. Write those formulas respectively as $\theta(\bar{a};y) \wedge \neg\phi(a_{i_0};y) \wedge \phi(a_{i_0+1};y)$ and $\theta(\bar{a};y) \wedge \phi(a_{i_0};y) \wedge \neg\phi(a_{i_0+1};y)$.

Increase the sequence $(a_i : i < \omega)$ to an indiscernible sequence $(a_i : i \in \mathbb{Q})$. Then for $i_0 \leq i, i' \leq i_0 + 1$, the formula $\theta(\bar{a};y) \wedge \phi(a_i;y) \wedge \neg\phi(a_{i'},y)$ is consistent if and only if $i < i'$. It follows that the formula $\psi(y;x) = \theta(\bar{a};y) \wedge \phi(x;y)$ has the strict order property. ⊣

Note that the formula we obtain in the proof has parameters. However it is clear from the definition of SOP that if the formula $\phi(x,y;d)$ has SOP, where d are parameters, then so does the formula $\phi(x;y\hat{\ }z)$.

2.3.4. Counting types. We give another characterization of NIP by counting types. This characterization however relies on set-theoretic assumptions, and will not be used later in this text.

Fix some formula $\phi(x;y)$. The *stability function* for ϕ is the function g_ϕ defined on cardinals by $g_\phi(\kappa) = \sup\{|S_\phi(A)| : A \text{ of size } \kappa\}$. Recall that $S_\phi(A)$ denotes the set of ϕ-types over A.

If $\phi(x;y)$ is stable, then for some polynomial $f(X)$, $g_\phi(\kappa)$ is bounded by $f(\kappa)$ for all finite or infinite κ. This follows from definability of types: Let n and z be as in Remark 2.57, then one can take $f(X) = nX^{|z|}$. Conversely, if $g_\phi(\kappa) = \kappa$ for some infinite cardinal κ, then ϕ is stable (Proposition 2.55).

If $\phi(x;y)$ has IP, it immediately follows from the definition that we have $g_\phi(\kappa) = 2^\kappa$ for every cardinal κ. And conversely if $g_\phi(\kappa) = 2^\kappa$ for every *finite* κ, then ϕ has IP.

The case of an unstable NIP formula $\phi(x;y)$ is trickier. We will see in Chapter 6, that there is a polynomial $f(X)$ such that for any *finite* cardinal κ, we have $g_\phi(\kappa) \leq f(\kappa)$. If κ is infinite and $2^\kappa = \kappa^+$, then as ϕ is unstable, we must have $g_\phi(\kappa) = 2^\kappa$ so there is no hope of separating IP from NIP by counting types over infinite sets without extra set-theoretic assumptions.

DEFINITION 2.68. For a cardinal λ, we define $\mathrm{ded}(\lambda) = \sup\{\kappa :$ there is a linear order of size κ which has a dense subset of size $\lambda\}$.

PROPOSITION 2.69. *Let $\phi(x;y)$ be a formula. Assume that there is some infinite set A with $|S_\phi(A)| > \mathrm{ded}(|A|)$, then $\phi(x;y)$ has IP.*

PROOF. Assume that $|S_\phi(A)| > \mathrm{ded}(|A|)$ and that A is chosen such that $\mu = |A|$ is minimal. Let $\lambda = \mathrm{ded}(|A|)^+$. Enumerate A as $\{a_i : i < \mu\}$ and for $i < \mu$, set $A_i = \{a_j : j < i\}$. For each $i < \mu$, we then have $|S_\phi(A_i)| \leq \mathrm{ded}(|A_i|) < \lambda$.

Define the following sets:

$$S_i = \{p \in S_\phi(A_i) : p \text{ has } \geq \lambda \text{ extensions to } S_\phi(A)\}, \text{ for } i < \mu;$$
$$S_\mu = \{p \in S_\phi(A) : p \upharpoonright A_i \in S_i \text{ for all } i < \mu\}.$$

Note that for every $i < \mu$, as $|S_\phi(A_i)| < \lambda$, we have $|S_\phi(A_i) \setminus S_i| < \lambda$ and as each type from that set has less than λ extensions to a type over A, the cardinality of $\{p \in S_\phi(A) : p \upharpoonright A_i \notin S_i\}$ is less that λ. Summing over $i < \mu$, we see that

$$|S_\phi(A) \setminus S_\mu| < \lambda.$$

It follows that every type in S_i has at least λ extensions to a type in S_μ.

Let $S_{<\mu} = \bigcup_{i<\mu} S_i$ and $S_{\leq\mu} = S_{<\mu} \cup S_\mu$. We define a linear order on $S_{\leq\mu}$ in the following way. For $p, q \in S_{\leq\mu}$, if $p \subseteq q$ (resp. $q \subseteq p$), we set $p \leq q$ (resp. $q \leq p$). Otherwise, let $i < \mu$ be maximal such that $p \upharpoonright S_i = q \upharpoonright S_i$. Then set $p < q$ if $p \vdash \neg\phi(x;a_i)$ (which implies $q \vdash \phi(x;a_i)$) and $p > q$ otherwise. We leave it to the reader to check that this indeed defines a linear order on $S_{\leq\mu}$ with $S_{<\mu}$ as a dense subset.

We now show:

(∗) For all $n < \omega$ and $q \in S_{<\mu}$, there are tuples $b_q^0, \dots, b_q^{n-1} \in A$ such that for every $\eta : n \to \{0,1\}$, the type $q(x) \wedge \bigwedge_{k<n} \phi(x;b_q^k)^{\eta(k)}$ is consistent.

This will imply that $\phi(x;y)$ has the independence property. We show (∗) by induction on n. For $n = 0$, there is nothing to prove.

Assume the result is known for n, and we prove it for $n+1$. Let $q \in S_i$ for some $i < \mu$. Define $S_q = \{p \in S_{\leq\mu} : p \upharpoonright S_i = q\}$. Using the order defined above, we see that $S_q \cap S_{<\mu}$ is a dense subset of S_q. We know that $|S_q| \geq \lambda$, therefore by definition of ded, $|S_q \cap S_{<\mu}| > \mu$. It follows that there is some $i < \mu$ such that $|S_q \cap S_i| > \mu$. The induction hypothesis gives, for every type $p \in S_q \cap S_i$, a tuple $(b_p^0, \dots, b_p^{n-1})$. We can find two distinct types $p_1, p_2 \in S_q \cap S_i$ for which the corresponding tuples are the same, equal to some $(b_0, \dots, b_{n-1})$. Let $b_n \in A_i$ be such that $p_1 \vdash \phi(x;b_n)$ and $p_2 \vdash \neg\phi(x;b_n)$ (exchanging the roles of p_1 and p_2 if necessary). Then for every $\eta : n+1 \to \{0,1\}$, the partial type $q(x) \wedge \bigwedge_{k\leq n} \phi(x;b_k)^{\eta(k)}$ is consistent. This finishes the induction step, and the proof. ⊣

For any infinite cardinal κ, we have $\kappa < \mathrm{ded}(\kappa) \leq 2^\kappa$ (Exercise 2.72). Hence if $2^\kappa = \kappa^+$, then $\mathrm{ded}(\kappa) = 2^\kappa$ and the hypothesis of the previous proposition cannot be satisfied. However, Mitchell has shown in [83] that if $\mathrm{cf}(\kappa) > \aleph_0$, then there is a cardinal-preserving forcing extension of the set-theoretic universe on which $\mathrm{ded}(\kappa) < 2^\kappa$. In a such an extension, a formula $\phi(x; y)$ is IP if and only if $g_\phi(\kappa) = 2^\kappa$.

2.3.5. More exercises.

Exercise 2.70. (T is NIP.) Let $M \models T$ and let p be a global M-invariant type. Let I be a Morley sequence of p over M. Assume that $p \upharpoonright MI$ is finitely satisfiable in M. Show that p is finitely satisfiable in M.

Exercise 2.71. (T is NIP.) Let I be an indiscernible sequence such that for any A over which I is based $\mathrm{Ev}(I/A)$ is definable. Show that I is totally indiscernible.

Exercise 2.72. Prove that for any cardinal κ, we have $\kappa < \mathrm{ded}(\kappa) \leq 2^\kappa$.

Exercise 2.73.
1. Let $\phi(x; y)$ be an NIP formula of VC-dimension d. Show that we have $\mathrm{alt}(\phi) \leq 2d$.
2. Give examples showing that one cannot conversely bound the VC-dimension in terms of the alternation rank.

Exercise 2.74. (T is NIP.) Let $p \in S(M)$ be a definable type. Show that p has a unique global coheir.

Exercise 2.75. (T is NIP.) Let p, q be two global M-invariant types. Assume that $p^{(\omega)}(\bar{x}) \otimes q^{(\omega)}(\bar{y})|_M = q^{(\omega)}(\bar{y}) \otimes p^{(\omega)}(\bar{x})|_M$. Show that $p(x) \otimes q(y) = q(y) \otimes p(x)$ (as global types).

References and related subjects

The definition of NIP and most results in Section 2.1 are due to Shelah and appeared first in [103] (see also [106, II.4]). Lemma 2.7 comes from Poizat [93]. The characterization 2.43 also appears in that paper, but the proof we give here is from [105]. Eventual types are defined in [94], although the terminology was introduced by Adler in [4]. Proposition 2.11 was first proved by Shelah using the $\mathrm{ded}(\lambda)$-characterization and the fact that the statement is absolute, see the discussion in [95, Chapter 12]. The proof we give here is very close to Poizat's approach in [95, 12.18], and was first published by Adler in [4].

The Baldwin-Saxl theorem is from [14]. See also Poizat [96]. Proposition 2.27 was proved by Shelah in [109].

Generically stable types are first defined by Shelah as 'stable types' in [107]. They are renamed and studied systematically by Hrushovski and Pillay in [61] and independently by Usvyatsov in [116]. This notion was extended outside of the NIP setting by Pillay and Tanović in [92].

Stability theory is an extremely rich subject on which a large number of papers have been written. We do not attempt to give an account of it here, but refer the reader to classical texts such as Tent and Ziegler [115], Marker [81], Poizat [95] for the basic theory, and Pillay [89], Baldwin [12] and Shelah [106] for more advanced material.

Directionality measures how many coheirs a type can have. Let $f(\lambda)$ be the maximal number of coheirs that a type over a model of size λ can have. In [68], Kaplan and Shelah show a trichotomy theorem for NIP theories: either f is absolutely bounded (*small directionality*), or it grows essentially like λ (*medium directionality*), or it grows like ded λ (*large directionality*). Examples of each type are given. Results about extracting indiscernible sequences in theories of small or medium directionality appear in Shelah [110].

Little is known in general about *Algebraic structures with NIP*. All known NIP fields are either algebraically closed, separably closed, real closed, or admit a non-trivial definable Henselian valuation. A full classification seems out of reach for now; in fact even the stable case is not known. In [66], Kaplan, Scanlon and Wagner show that NIP fields have finitely many Artin-Schreier extensions, along with results about valued fields.

Some theorems about definable groups related to the Baldwin-Saxl theorem will be presented in Chapter 8. A systematic study of such 'chain conditions' can be found in Kaplan and Shelah [69]. Most results there require stronger conditions than NIP (in particular *strongly dependent* which we will define in Chapter 4). In a completely different direction, Macpherson and Tent have studied pseudofinite NIP groups in [80].

Chapter 3

HONEST DEFINITIONS AND APPLICATIONS

A fundamental characteristic property of stable theories is *definability of types*, namely the property that if $A \subset \mathcal{U}$ is any subset (big or small), and $\phi(x; b) \in L(\mathcal{U})$, a formula, then the set $\phi(A; b)$ coincides with the trace on A of some A-definable set: there is $\psi(x; d) \in L(A)$ such that $\phi(A; b) = \psi(A; d)$. In other words, we can internalize the parameters of ϕ inside A (up to changing the formula). In this chapter, we show that a weak form of definability of types holds in NIP theories. We do not manage to find a definition of $\phi(A; b)$ with parameters inside A, but we do with parameters in some elementary extension A' of A. Furthermore, the definition satisfies a property called *honesty* which says that, on A', the new formula lies inside the original one.

Associated with definability of types are so-called *reflection principles* stating that if some (e.g., type-definable) set A is internally simple in some sense, then it is also externally simple: its interactions with the rest of the structure cannot be too complicated. We obtain some statements of this kind using honest definitions. Note that the characteristic property 2.7 of NIP is an example of such a phenomenon. It says that if a sequence I is indiscernible (internal simplicity), then the intersection of I with some unary $\mathcal{U}$-definable set is a finite union of convex subsets of I (external simplicity). We will in fact extend this result to definable sets of higher arity and conclude that when we add parameters to the base, we can shrink an indiscernible sequence to a subsequence which remains indiscernible over the additional parameters.

3.1. Stable embeddedness and induced structure

3.1.1. Stable embeddedness.

Definition 3.1. Let $\pi(x)$ be a partial unary type over $\emptyset$. We say that $\pi(x)$ is *stably embedded* if for every formula $\phi(x_1, \dots, x_n; b)$, $b \in \mathcal{U}$, there is a formula $\psi(x_1, \dots, x_n; z)$ and $d \in \pi(\mathcal{U})$ such that $\phi(x_1, \dots, x_n; b)$ and $\psi(x_1, \dots, x_n; d)$ agree on tuples of realizations of $\pi(x)$.

The following observation is a standard compactness argument.

OBSERVATION 3.2. *If $\pi(x)$ is stably embedded, then one can choose the formula $\psi(x_1, \ldots, x_n; z)$ in a way that it depends only on $\phi(x_1, \ldots, x_n; y)$ and not on the parameters b.*

It will be convenient for us to generalize the definition of stable embeddedness to arbitrary subsets of $\mathcal{U}$. However, we have to be careful as the previous observation may not hold any more, and thus we will distinguish between weak and uniform stable embeddedness.

DEFINITION 3.3. Let $A \subseteq \mathcal{U}$ be any subset (big or small). We say that A is *weakly stably embedded* (in $\mathcal{U}$) if given any formula $\phi(x_1, \ldots, x_n; y)$ and $b \in \mathcal{U}$, there is some $\psi(x_1, \ldots, x_n; z)$ and $d \in A$ such that for every $a \in A^n$:

$$\models \phi(a; b) \iff \models \psi(a; d).$$

We say that A is *uniformly stably embedded* (or just *stably embedded*) if the formula $\psi(x_1, \ldots, x_n; z)$ depends only on $\phi(x_1, \ldots, x_n; y)$ and not on the parameters b.

By A being "big or small", we mean that there is no cardinality restriction on A: it could be a small subset, a definable set, a type-definable set etc.

Example 3.4. Let $M = (\mathbb{R}; \leq)$, seen as a model of DLO. Then M is (uniformly) stably embedded in $\mathcal{U}$.

More generally, a small set A is weakly stably embedded in $\mathcal{U}$ if and only if all types over A are definable. The theory T is stable if and only if any set A is weakly stably embedded if and only if any set A is stably embedded.

Example 3.5. In the theory ACVF of algebraically closed valued fields, both the residue field and the value group are stably embedded. (See Section A.2.)

3.1.2. Induced structure. Let $A \subseteq \mathcal{U}$ be a set big or small (as above, the relevant cases are: A is a small set, A is a definable set or A is a type-definable set). We would like to consider A as a structure in its own right.

DEFINITION 3.6. Let $A \subseteq \mathcal{U}$ be any subset and let $B \subset \mathcal{U}$ be a small set of parameters. We define $A_{\text{ind}(B)}$ to be the structure in the language $L_B = \{R_{\phi(\bar{x})}(\bar{x}) : \phi(\bar{x}) \in L(B)\}$ whose universe is A and where each $R_\phi(\bar{x})$ is interpreted the obvious way: for every $\bar{a} \in A$, $A_{\text{ind}(B)} \models R_\phi(\bar{a}) \iff \mathcal{U} \models \phi(\bar{a})$.

For an arbitrary set A, the move from seeing A as a subset of $\mathcal{U}$ to seeing it as an independent structure is often unnatural. In particular, even if A is stably-embedded in $\mathcal{U}$, the structure $A_{\text{ind}(\emptyset)}$ may be intractable because of quantifiers now ranging over A. In general, the structure $A_{\text{ind}(B)}$ is meaningful only if A has some extra properties (notably if A is a model or a definable set) or if we only pay attention to quantifier-free formulas.

Example 3.7. If $D = \phi(\mathcal{U})$ is a B-definable set, then the structure $D_{ind(B)}$ eliminates quantifiers: The formula $\exists x(R_\psi(x, y))$ is equivalent to $R_{\psi'}(y)$ where $\psi'(y) = \exists x(\phi(x) \wedge \psi(x; y))$. In particular $D_{ind(B)}$ is NIP (resp. stable) if T is.

If D is a $\emptyset$-definable set, then D is stably embedded if and only if $D_{ind(\emptyset)}$ has the same definable sets as $D_{ind(B)}$, for any $B \subset \mathcal{U}$.

If $M \prec \mathcal{U}$ is a model, then $M_{ind(\emptyset)}$ is just an expansion by definitions of the structure M. However, if we replace $\emptyset$ by some $B \subset \mathcal{U}$ not included in M, we might get new definable sets.

DEFINITION 3.8. An *externally definable* subset of M is a subset $D \subseteq M^k$ of the form $\phi(M; b)$ for some $b \in \mathcal{U}$.

The formula $\phi(x; b)$ is called an *external definition* of D.

In a stable theory, any externally definable subset of a model M is actually definable. An important intuition of Shelah is that in NIP theories, externally definable sets are well-behaved and one should consider them alongside definable sets. Contemplate for example the difference between the following two situations:

- Let (M, R) be a model of the random graph. Then any subset of M (in dimension 1) is externally definable.
- Let $(M, \leq)$ be a model of DLO, then externally definable sets in dimension 1 are exactly the finite unions of convex subsets of M.

DEFINITION 3.9. Let $M \prec N$, N is $|M|^+$-saturated. Then $M_{ind(N)}$ is called the *Shelah expansion* of M and is denoted by M^{Sh}.

Formally, M^{Sh} depends on the choice of N: taking different N yields structures in different languages. However all those structures have the same definable sets. This justifies talking about *the* Shelah expansion. Note also that in M^{Sh} all elements of M are in $\mathrm{dcl}(\emptyset)$ since any singleton is a definable set.

We will say more in Section 3.3 about the Shelah expansion of an NIP structure.

Our goal in this chapter is to understand the quantifier-free structure of $A_{\mathrm{ind}(B)}$ in terms of that of $A_{\mathrm{ind}(\emptyset)}$. As we have seen, if T is stable, then A is stably embedded and the two structures are essentially equal. If T is NIP we will show that the quantifier-free definable sets of $A_{\mathrm{ind}(B)}$ are quantifier-free definable in the Shelah expansion of $A_{\mathrm{ind}(\emptyset)}$.

Exercise 3.10. Let $M \prec \mathcal{U}$ and let $\hat{N}$ be an elementary extension of M^{Sh} with L-reduct $N \prec \mathcal{U}$. Show that up to a renaming of the language, $\hat{N}$ is equal to $N_{ind(B)}$ for some set $B \subset \mathcal{U}$ of parameters.

Observe that in general $\hat{N}$ is a proper reduct of N^{Sh}.

3.1.3. Pairs. Another way to single out a subset A from $\mathcal{U}$ is to name the set A by a new unary predicate. This creates what we will call a pair.

More precisely, let M be an L structure, $A \subseteq M$. Let $L_P = L \cup \{\boldsymbol{P}(x)\}$, where $\boldsymbol{P}(x)$ is a new unary relation symbol. We will sometimes write $x \in \boldsymbol{P}$ instead of $\boldsymbol{P}(x)$. The *pair* (M, A) is the L_P-structure whose L-reduct is M, and where $\boldsymbol{P}$ is interpreted by: $\boldsymbol{P}(a) \iff a \in A$.

The structure (M, A) is richer than M and $A_{\text{ind}(M)}$ in the sense that both are interpretable in it. In many cases, one actually seeks to understand the pair (M, A) in terms of the two simpler structures M and $A_{\text{ind}(M)}$ (or even $A_{\text{ind}(\emptyset)}$). However, for us, the pair construction will be just a technical tool and we will not study the properties of the pair *per se*.

Exercise 3.11. Let $A \subseteq M$, where M is $|A|^+$-saturated. Then A is uniformly stably embedded if and only if for any elementary extension (M', A') of the pair (M, A) and any $m \in M'$, $\text{tp}_L(m/A')$ is definable.

Exercise 3.12. Let $I \subset M$ be an indiscernible sequence. Let (M', I') be an elementary extension of the pair (M, I). Then there is an ordering on I' making it into an indiscernible sequence.

3.2. Honest definitions

We now state the main theorem of this chapter.

THEOREM 3.13. *Let $M \models T$, $A \subseteq M$, $\phi(x; y) \in L$ and $b \in M$ a $|y|$-tuple. Assume that $\phi(x; y)$ is NIP. Then there is an elementary extension $(M, A) \prec (M', A')$, a formula $\psi(x; z) \in L$ and a tuple d of elements of A' such that*

$$\phi(A; b) \subseteq \psi(A'; d) \subseteq \phi(A'; b).$$

Remark 3.14. The conclusion of the theorem is equivalent to saying that there is a formula $\psi(x; z)$ such that for any finite $A_0 \subseteq \phi(A; b)$, there is a tuple d of elements of A such that $A_0 \subseteq \psi(A; d) \subseteq \phi(A; b)$.

Indeed, if this is true, then the theorem follows by compactness in the structure (M, A). Conversely, if the conclusion of the theorem holds and $A_0 \subseteq A$ is finite, then the formula $\psi(x; d)$ given by the theorem satisfies $A_0 \subseteq \psi(A'; d)$ and $(M', A') \models \forall x \in \boldsymbol{P}(\psi(x; d) \to \phi(x; b))$. By elementarity of the extension $(M, A) \prec (M', A')$, we can find some $d' \in A$ such that $\psi(x; d')$ has the same two properties.

Example 3.15. Take T to be DLO, $M = (\mathbb{R}, <)$ and $A = \mathbb{Q}$. Let $\phi(x; b)$ be $x \leq b$, where $b = \sqrt{2}$, say, and let $\psi(x; z) = x \leq z$. We see that for any finite $A_0 \subset \phi(A; b)$, there is some point $d \in A$ such that $A_0 \subseteq \psi(A; d) \subseteq \phi(A; b)$: Take d to be a rational smaller than $\sqrt{2}$, but greater than all the elements of A_0.

PROOF OF THEOREM 3.13. Let $S_A \subset S_x(\mathcal{U})$ be the set of global types (in the variable x), finitely satisfiable in A. It is a closed subset of $S_x(\mathcal{U})$, and therefore compact. Let $(M', A') \succ (M, A)$ be a $|M|^+$-saturated elementary extension. Let $p \in S_A$. We try to build a sequence $(a_i : i < \omega)$ such that for all i we have:

- $a_i \in A'$;
- $a_i \models p \upharpoonright Aa_{<i}$;
- $\models \neg(\phi(a_i; b) \leftrightarrow \phi(a_{i+1}; b))$.

If we succeed, then the sequence $(a_i : i < \omega)$ is a Morley sequence of p over A and as such it is L-indiscernible. Then the third condition implies that ϕ has infinite alternation rank, contradicting NIP. We conclude that the construction must stop at some finite stage. So assume we have built $(a_i : i < n)$, and we cannot find a point a_n. Let $\varepsilon_p \in \{0, 1\}$ be such that $p \vdash \phi(x; b)^{\varepsilon_p}$. We first argue that $\phi(a_{n-1}; b)^{\varepsilon_p}$ holds. The type p is finitely satisfiable in A, therefore for every subset $B \subset M'$ of size $\leq |M|$, the L_P-type $p|_B(x) \cup \{\boldsymbol{P}(x)\}$ is finitely satisfiable in A and thus realized in A'. Taking $B = Aba_{<n}$ we see that if we had $\models \neg\phi(a_{n-1}; b)^{\varepsilon_p}$, then we could take a_n realizing $p|_B(x) \cup \{\boldsymbol{P}(x)\}$ and obtain an extra alternation.

By compactness, there is a formula $\theta_p(x) \in L(Aa_{<n})$ such that $p \vdash \theta_p(x)$ and $(M', A') \models \forall x \in \boldsymbol{P}(\theta_p(x) \to \phi(x; b)^{\varepsilon_p})$. As S_A is compact, we can find a finite set $S_0 \subset S_A$ such that $\bigcup_{p \in S_0} \theta_p(x)$ covers S_A. Set

$$\psi(x) = \bigvee_{\substack{p \in S_0 \\ \varepsilon_p = 1}} \theta_p(x).$$

Write $\psi(x) = \psi(x; d)$ where $d \in A'$ are the parameters appearing in ψ. We show that ψ has the required properties. First, for all types $p \in S_A$ such that $p \vdash \phi(x; b)$, we have $p \vdash \psi(x; d)$. In particular, this is true for realized types. It follows that $\phi(A; b) \subseteq \psi(A; d) \subseteq \psi(A'; d)$. Furthermore, we have $(M', A') \models \forall x \in \boldsymbol{P}(\psi(x; d) \to \phi(x; b))$, therefore $\psi(A', d) \subseteq \phi(A', b)$. $\dashv$

DEFINITION 3.16. We say that the formula $\psi(x; d)$ in the previous theorem is an *honest definition* of $\phi(x; b)$ over A.

The following corollary is simply the weakening of the theorem obtained by removing in the conclusion the "honesty" hypothesis $\psi(A'; d) \subseteq \phi(A'; b)$. We state it separately because it might seem more natural and because it is sufficient for some applications.

COROLLARY 3.17. *Let $M \models T$, $A \subset M$, $\phi(x; y) \in L$ and $b \in M^{|y|}$. Assume that $\phi(x; y)$ is NIP. Then there is an elementary extension $(M, A) \prec (M', A')$, a formula $\psi(x; z) \in L$ and a $|z|$-tuple $d \in A'$ such that*

$$\phi(A; b) = \psi(A; d).$$

COROLLARY 3.18. (T *is NIP.*) *Let $M \models T$ and $A \subseteq M$. Let $b \in M$ be a finite tuple. Let $(M, A) \prec (M', A')$ be an $|M|^+$-saturated extension. Then there* *is $A_0 \subseteq A'$ of size at most $|T|$ such that for any two tuples a and a' from A we have:*

$$a \equiv_{A_0} a' \Longrightarrow a \equiv_b a'.$$

PROOF. For any formula $\phi(x; y)$, $|y| = |b|$, let $\psi_\phi(x; d_\phi)$ be an honest definition of $\phi(x; b)$ over A where $d_\phi \in A'$. Take A_0 to contain the union of the parameters d_ϕ for ϕ ranging over all formulas. ⊣

PROPOSITION 3.19. *(T is NIP.) Let $A \subseteq M$. Assume that in $A_{\text{ind}(\emptyset)}$, the quantifier-free formulas are stable. Then A is (uniformly) stably embedded.*

PROOF. Let $\phi(x; b)$ be a formula with parameters in M. By Corollary 3.17, there is $(M', A') \succ (M, A)$ and $\psi(x; d) \in L(A')$ such that $\phi(A; b) = \psi(A; d)$. Note that we have $A_{\text{ind}(\emptyset)} \prec A'_{ind(\emptyset)}$ and $\psi(A; d) = R_{\psi(x;z)}(A; d)$. We work inside the structure $A'_{ind(\emptyset)}$. The formula $R_{\psi(x;z)}(x; z)$ is stable and by Proposition 2.58 the set $R_{\psi(x;z)}(A; d)$ is definable by a Boolean combination of formulas of the form $R_\psi(x; c)$, $c \in A$. This translates into some formula $\theta(x; \bar{c}) \in L(A)$ such that $\theta(A; \bar{c}) = \phi(A; d)$ as required. Uniformity follows from uniformity in 2.58. ⊣

Exercise 3.20. (T is NIP.) Let $D \subseteq M$ be any subset. Let $\phi(x, y)$ be a $\emptyset$-definable formula which defines a total order $<$ on D. Assume that $D_{ind(\emptyset)}$ is o-minimal when equipped with that ordering. Then any subset of D definable in M is a finite union of $<$-convex sets in D.

3.3. Naming a submodel

Assumption: Throughout this section, we assume that T is NIP.

PROPOSITION 3.21. *Let $D \subseteq M^k$ be an externally definable set. Then there is an external definition $\phi(x; b) \in L(\mathcal{U})$ of D with the following property:*

(∗) *For every formula $\theta(x; a) \in L(M)$, $|x| = k$, such that $D \subseteq \theta(M; a)$, we have $\mathcal{U} \models \phi(x; b) \to \theta(x; a)$.*

PROOF. Let $M \prec N$ and $\phi_0(x; b_0) \in L(N)$ an external definition of D. Theorem 3.13 applied to the pair (N, M) and the formula $\phi_0(x; b_0)$ yields an elementary extension $(N, M) \prec (N', M')$ and a formula $\phi(x; b) \in L(M')$ such that $D = \phi(M; b)$ and $\phi(M'; b) \subseteq \phi_0(M'; b_0)$. Let $\theta(x; a) \in L(M)$ be a formula such that $D \subseteq \theta(M; a)$. Then $(N, M) \models \forall x \in \boldsymbol{P}(\phi_0(x; b_0) \to \theta(x; a))$. Therefore the same sentence holds in the pair (N', M'). We conclude that $\phi(M'; b) \subseteq \phi_0(M'; b_0) \subseteq \theta(M'; a)$. As M' is a model and $b \in M'$, we have $M' \models \phi(x; b) \to \theta(x; a)$ as required. ⊣

Example 3.22. Let T be DLO, and let $M \models T$. Let $a \in M$ and let $b \in \mathcal{U}$ such that $b > a$, but $b < m$ for any $a < m \in M$. Consider the formula $\phi(x; b) = x > b$. Then $\phi(M; b)$ coincides with the definable set $x > a$. The formula $\phi(x;b)$ has the property $(*)$ of the previous proposition. However the formula $\neg\phi(x;b)$ which defines on M the set $x \leq a$ does not. Indeed letting $\theta(x;a) = x \leq a$, we have $\neg\phi(M;b) \subseteq \theta(M;a)$, but $\mathcal{U} \not\models \neg\phi(x;b) \to \theta(x;a)$.

Recall the construction of the Shelah expansion presented at the beginning of this chapter. We will now show one way in which externally definable sets are *well behaved* in NIP theories.

PROPOSITION 3.23. *The structure M^{Sh} admits elimination of quantifiers.*

PROOF. We have to show that the projection of an externally definable set is again externally definable. Let $D \subseteq M^{k_1+k_2}$ be externally definable. Let $\phi(x_1, x_2; b) \in L(\mathcal{U})$ be an external definition of D given by Proposition 3.21 (where $|x_1| = k_1$ and $|x_2| = k_2$). Let π denote the projection $M^{k_1+k_2} \to M^{k_1}$ and set $\psi(x_1; b) = \exists x_2 \phi(x_1, x_2; b)$. We claim that $\pi(D) = \psi(M; b)$.

It is clear that $\pi(D) \subseteq \psi(M;b)$. To show the other inclusion, take some $a \in M^{k_1} \setminus \pi(D)$. Let $\zeta(x_1, x_2) = (x_1 \neq a)$. Then $D \subseteq \zeta(M)$. By hypothesis on $\phi(x_1, x_2; b)$, we have $\models \phi(x_1, x_2; b) \to \zeta(x_1, x_2)$, therefore $\models \psi(x_1; b) \to x_1 \neq a$. It follows that $a \notin \psi(M;b)$ as required. ⊣

It follows that we have $(M^{Sh})^{Sh} \cong M^{Sh}$, in the sense that the former is an expansion by definition of the latter. In other words, M^{Sh} is weakly stably embedded in a monster model of its theory. However, it is not in general uniformly stably embedded.

COROLLARY 3.24. *The structure M^{Sh} is NIP.*

PROOF. This follows easily from Proposition 3.23: Let $N \succ M$ be $|M|^+$-saturated and construct M^{Sh} using the model N so that M^{Sh} has quantifier elimination in the language $\{R_\phi(x;y) : \phi \in L(N)\}$. By Lemma 2.9, we only need to check that each formula $R_\phi(x;y)$ is NIP, but such a formula cannot have alternation rank greater than that of $\phi(x;y)$. ⊣

Remark 3.25. There are structures with IP which satisfy Proposition 3.23. For example if M is a model of arithmetic, than any subset of any M^k is externally definable.

One might wonder what happens if instead of naming all externally definable sets of M, we name for example only those definable from a given tuple in $\mathcal{U}$. It turns out that quantifier elimination may fail, as the following example shows.

Example 3.26. Let $(\mathbb{Q}; 0, +, <)$ be the additive group of rational numbers; a model of the complete theory of ordered $\mathbb{Q}$-vector spaces. It is an o-minimal structure. In an elementary extension, find some $\varepsilon > 0$ which is infinitesimal and let $M = \mathbb{Q}(\varepsilon)$; the vector space generated by $\mathbb{Q}$ and ε. Let $p \in S_1(M)$ be a type of an element falling into the cut corresponding to an irrational number (say $\sqrt{2}$) and take $a \models p$. We show that $M_{ind(Ma)}$ does not admit elimination of quantifiers.

Consider the set $D = \{b \in M : \neg(\exists c_1, c_2 \in M)(c_1 < a \wedge c_2 > a \wedge b = c_2 - c_1)\}$. Then D is definable in $M_{ind(Ma)}$ and corresponds to the set of infinitesimal elements. However, D is not definable without quantifiers (because of o-minimality and the fact that there is no new infinitesimal in $\mathbb{Q}(\varepsilon, a)$).

It is an open problem to find general sufficient conditions which ensure that the structure $M_{ind(A)}$ eliminates quantifiers. In particular it is not known whether $M_{ind(I)}$ eliminates quantifiers when I is a Morley sequence of an M-invariant type p. A partial result is obtained in [114, Theorem 3.36] where a positive answer is obtained when $p^{(\omega)}$ is both an heir and a coheir of its restriction to M.

Exercise 3.27. Show that Proposition 3.23 fails for the random graph.

Exercise 3.28. Show that there is a canonical homeomorphism between the space of types over $\emptyset$ in the structure M^{Sh} and the set of L-types over $\mathcal{U}$ which are finitely satisfiable in M (seen as a subspace of $S(\mathcal{U})$).

Exercise 3.29. Let $M \models T$, $\phi(x, y; b) \in L(\mathcal{U})$. Assume that $\phi(M; b)$ is the graph of a function. Then there is a formula $\psi(x, y; d) \in L(\mathcal{U})$ such that $\phi(M; b) = \psi(M; d)$ and $\mathcal{U} \models (\forall x)(\exists^{\leq 1} y)\psi(x, y; d)$ (so $\psi(\mathcal{U}; d)$ is the graph of a partial function).

3.4. Shrinking of indiscernible sequences

Assumption: Throughout this section, we assume that T is NIP.

We are interested now in the case where $A = I$ is an indiscernible sequence. We study the trace on I^n of definable sets in n variables. The case $n = 1$ has already been dealt with in Chapter 2: the trace on I of a definable set is a finite union of $<_I$-convex sets.

In particular, given some tuple b of parameters and an indiscernible sequence I, we are interested in ways to shrink the sequence I to a subsequence I' which is indiscernible over b. We have already seen during the proof of Proposition 2.11 that if the sequence I has large cofinality, then some end segment of it is indiscernible over b. We will generalize this result by showing that I can be broken into at most $2^{|T|+|b|}$ segments, each indiscernible over b. Moreover, this remains true working over any base set of parameters or with a sequence of infinite tuples.

We will give several statements. Later results generalize previous ones, so there is some redundancy.

DEFINITION 3.30. Let $(\mathcal{I}, \leq)$ be a linear order. A *convex equivalence relation* $\sim$ on $\mathcal{I}$ is an equivalence relation all of whose classes are convex sets. The equivalence relation $\sim$ is said to be *finite* if it has finitely many classes.

If $\sim$ is a convex equivalence relation on $(\mathfrak{I}, \leq)$, we extend $\sim$ to cartesian powers of $\mathfrak{I}$: if $\bar{i} = (i_1, \ldots, i_n), \bar{j} = (j_1, \ldots, j_n) \in \mathfrak{I}^n$, then we set $\bar{i} \sim \bar{j}$ if the tuples $\bar{i}$ and $\bar{j}$ have the quantifier-free type in the structure $(\mathfrak{I}; \leq)$ and $i_k \sim j_k$ for all k.

For any family $\bar{c}$ of elements of $\mathfrak{I}$, we define the convex equivalence relation $\sim_{\bar{c}}$ by $i \sim_{\bar{c}} j$ if i and j have the same quantifier-free type in the structure $(\mathfrak{I}; \leq, \bar{c})$ (elements of $\bar{c}$ are named by constants).

We say that a convex equivalence relation $\sim$ is *essentially of size* κ if it is the intersection of κ many finite convex equivalence relations. Note that a relation essentially of size κ has at most 2^κ many classes.

Remark 3.31. If $\mathfrak{I}$ is a linear order, let $\mathfrak{J} = \mathrm{compl}(\mathfrak{I})$ be the completion of $\mathfrak{I}$. Let $\sim$ be any convex equivalence relation on $\mathfrak{I}$ essentially of size κ. Then there is $\bar{c} \subseteq \mathfrak{J}$ of size $\leq \kappa$ such that the restriction of $\sim_{\bar{c}}$ to $\mathfrak{I}$ refines $\sim$ and for every $i, j \in \mathfrak{I} \setminus \bar{c}$, we have $i \sim j \iff i \sim_{\bar{c}} j$.

If $\sim$ is finite, then $\bar{c}$ can be taken to be finite.

PROPOSITION 3.32. *Let $I = (a_i : i \in \mathfrak{I})$ be an indiscernible sequence. Let $\phi(x_1, \ldots, x_n; b) \in L(\mathcal{U})$ be a formula. Then there is a finite convex equivalence relation $\sim$ on $\mathfrak{I}$ such that for all $\bar{i}, \bar{j} \in \mathfrak{I}^n$, we have*

$$\bar{i} \sim \bar{j} \Rightarrow \phi(a_{\bar{i}}; b) \leftrightarrow \phi(a_{\bar{j}}; b).$$

PROOF. Without loss of generality, and for simplicity of notations, we assume that I is a sequence of singletons (since we can work in T^{eq} for example).

Let M be a model containing I and b. We expand the pair (M, I) by adding a binary predicate $\boldsymbol{E}(x, y)$ interpreted as $\boldsymbol{E}(M) = \{(a_i, a_j) : i \leq_{\mathfrak{I}} j\}$. Let (M', I') be an elementary extension of (M, I) which is $|M|^+$-saturated. Then we can write $I' = (a_i : i \in \mathfrak{I}')$ where $\mathfrak{I}' \supset \mathfrak{I}$ is ordered such that $i <_{\mathfrak{I}'} j$ if and only if $M' \models \boldsymbol{E}(a_i, a_j)$. By elementarity of the extension, I' is an L-indiscernible sequence.

By Corollary 3.17, there is a formula $\psi(x_1, \ldots, x_n; d) \in L(I')$ such that $\phi(I; b) = \psi(I; d)$. Write $d = (a_{c_1}, \ldots, a_{c_k})$ and $\bar{c} = (c_1, \ldots, c_k)$. As I' is indiscernible, the truth value of $\psi(a_{\bar{i}}; d)$ depends only on the quantifier-free type $\mathrm{qftp}(\bar{i}/\bar{c})$ in $(\mathfrak{I}', \leq)$, hence on the class of $\bar{i}$ modulo $\sim_{\bar{c}}$. Now define $\sim$ to be the restriction of $\sim_{\bar{c}}$ to $\mathfrak{I}$. ⊣

THEOREM 3.33. *Let A be a small set of parameters and $I = (a_i : i \in \mathfrak{I})$ an A-indiscernible sequence. Let $\phi(\bar{x}, y; b) \in L(\mathcal{U})$ with $\bar{x} = (x_1, \ldots, x_n)$. Then there is a finite convex equivalence relation $\sim$ on $\mathfrak{I}$ such that for $\bar{i}, \bar{j} \in \mathfrak{I}^n$, we have*

$$\bar{i} \sim \bar{j} \Rightarrow \forall e \in A, \phi(a_{\bar{i}}, e; b) \leftrightarrow \phi(a_{\bar{j}}, e; b).$$

PROOF. The proof is similar to that of Proposition 3.32, except that instead of naming the sequence I, we name the product $A \times I$ (as a subset of M^{eq}).

Take M a model containing A, I and b, and consider the pair $(M, A \times I)$. By Corollary 3.17, there is an elementary extension $(M, A \times I) \prec (M', A' \times I')$

and a formula $\psi(x_1, \ldots, x_n, y; a, d)$, $a \in A'$, $d \in I'$ such that $\phi(I^n \times A; b) = \psi(I^n \times A; a, d)$. By the same argument as in the previous proposition, we may assume that $I' = (a_i : i \in \mathfrak{I}')$ with $\mathfrak{I}' \supset \mathfrak{I}$ is L-indiscernible over A'. Write $d = (a_{c_1}, \ldots, a_{c_k})$. One then checks that the relation $\sim$ on $\mathfrak{I}$ defined as the restriction of $\sim_{\bar{c}}$ to $\mathfrak{I}'$ has the required property. ⊣

Remark 3.34. It follows from the theorem and Lemma 3.31 that if I is ordered by a complete order and if there is a formula $\theta(x, y) \in L(I)$ which orders I, then I is stably embedded.

(Even if I is not totally indiscernible, such a formula $\theta(x, y)$ may not exist. Take for example a circular order $(M; R(x, y, z))$, where $R(x, y, z)$ is the betweenness relation and an indiscernible sequence $I = (a_i : i \in \mathbb{Q})$ of points from it. Then the involution $a_k \mapsto a_{-k}$ is elementary, so we cannot define the order on I without extra parameters. This is the only possible obstruction; see [57].)

COROLLARY 3.35. *Let A be a small set of parameters and $I = (a_i : i \in \mathfrak{I})$ an A-indiscernible sequence of finite tuples. Let $b \in \mathfrak{U}$ be a finite tuple. Then there is a convex equivalence relation $\sim$ on $\mathfrak{I}$ essentially of size $|T|$ such that for any $\phi(\bar{x}, y; z) \in L$ with $\bar{x} = (x_1, \ldots, x_n)$ for any $\bar{i}, \bar{j} \in \mathfrak{I}^n$, we have*

$$\bar{i} \sim \bar{j} \Longrightarrow \forall e \in A, \phi(a_{\bar{i}}, e; b) \leftrightarrow \phi(a_{\bar{j}}, e; b).$$

We obtain what we announced in the introduction of this section: for any (finite or infinite) tuple b and A-indiscernible sequence I, the sequence I can be broken into at most $2^{|T|+|b|}$ convex segments, each indiscernible over Ab. To see this, let $\sim$ be the intersection of all the convex equivalence relations given by Corollary 3.35 for finite subtuples of b and break I into $\sim$-equivalence classes.

It turns out that this result is also true when the sequence I is composed of tuples of infinite length, although we need to redo the proof in that case.

THEOREM 3.36. *Let A be a small set of parameters and $I = (\bar{a}_i : i \in \mathfrak{I})$ an A-indiscernible sequence of tuples of arbitrary length. Let $b \in \mathfrak{U}$ be a finite tuple. Then there is a convex equivalence relation $\sim$ on $\mathfrak{I}$ essentially of size $|T|$ such that for any $\phi(\bar{x}, y; z) \in L$ with $\bar{x} = (\bar{x}_1, \ldots, \bar{x}_n)$ for any $\bar{i}, \bar{j} \in \mathfrak{I}^n$, we have*

$$\bar{i} \sim \bar{j} \Longrightarrow \forall e \in A, \phi(\bar{a}_{\bar{i}}, e; b) \leftrightarrow \phi(\bar{a}_{\bar{j}}, e; b).$$

PROOF. Let M be a model containing A, I and b. Write $\bar{a}_i = (a_i^j : j < \alpha)$. Let $\boldsymbol{P}(x)$ and $\boldsymbol{A}(x)$ be two new unary predicates, and $\boldsymbol{E}(x, y)$, $\boldsymbol{F}(x, y)$, $\boldsymbol{R}(x, y)$ be new binary predicates. Set $L' = L \cup \{\boldsymbol{P}, \boldsymbol{A}, \boldsymbol{E}, \boldsymbol{F}, \boldsymbol{R}\}$ and expand M into an L'-structure $(M; P, A, E, F, R)$ by setting $\boldsymbol{P}(M) = \{a_i^j : j < \alpha, i \in \mathfrak{I}\} \cup A$, $\boldsymbol{A}(M) = A$, $\boldsymbol{E}(M) = \{(a_i^j, a_i^{j'}) : i \in \mathfrak{I}, j, j' < \alpha\}$, $\boldsymbol{F}(M) = \{(a_i^j, a_{i'}^j) : i, i' \in \mathfrak{I}, j < \alpha\}$ and $\boldsymbol{R}(M) = \{(a_i^j, a_{i'}^{j'}) : i < i' \in \mathfrak{I}, j, j' < \alpha\}$.

Let $(M; P, A, E, F, R) \prec (M'; P', A', E', F', R')$ be some $|M|^+$-saturated elementary extension. Using the extra structure given by $\boldsymbol{E}$, $\boldsymbol{F}$ and $\boldsymbol{R}$, we can write $P' = A' \cup \{a_i^j : i \in \mathfrak{I}', j < \beta\}$ for some $\beta \geq \alpha$ and $\mathfrak{I}' \supset \mathfrak{I}$ such that the sequence $((a_i^j)_{j<\beta} : i \in \mathfrak{I}')$ is indiscernible over A'. By Corollary 3.18, there is some set $P_0 \subset P'$ of size at most $|T|$ such that for any two tuples $a, a' \in P$, we have

$$a \equiv_{P_0} a' \Longrightarrow a \equiv_b a'.$$

Let $\bar{c} \subset \mathfrak{I}'$ be the family of elements $c \in \mathfrak{I}'$ for which there is $j < \beta$ such that $a_c^j \in P_0$. The equivalence relation $\sim_{\bar{c}}$ on $\mathfrak{I}'$ is a convex equivalence relation essentially of size $|T|$. Let $\sim$ be its restriction to $\mathfrak{I}$. We check that $\sim$ has the required property: let $\bar{i}, \bar{j} \in \mathfrak{I}^n$, be $\sim$-equivalent and let $e \in A$. Set $a = a_{\bar{i}}\hat{\ }e$ and $a' = a_{\bar{j}}\hat{\ }e$. Then as the sequence $(\bar{a}_i : i \in \mathfrak{I}')$ is indiscernible over e, we have $a \equiv_{P_0} a'$ and hence $a \equiv_b a'$ as required. ⊣

Exercise 3.37 (Critical points). Let $I = (a_i : i \in \mathfrak{I})$ be an A-indiscernible sequence of finite tuples. Assume that the order $\mathfrak{I}$ is dense complete without end points. Let $b \in \mathcal{U}$ be a finite tuple. Let $\phi(x_1, \ldots, x_n, y; z)$ be a formula with $|x_k| = |a_i|$ and $|z| = |b|$. Call an index $i \in \mathfrak{I}$ *ϕ-critical* if there is $k \in \{1, \ldots, n\}$, some $e \in A$ and indices $i_1 < \cdots < i_n$ in $\mathfrak{I}$ such that in any open interval around i we can find i'_k, i''_k satisfying:

$$\models \phi(a_{i_1}, \ldots, a_{i'_k}, \ldots, a_{i_n}, e; b) \wedge \neg\phi(a_{i_1}, \ldots, a_{i''_k}, \ldots, a_{i_n}, e; b).$$

Let $\bar{c}$ denote the set of ϕ-critical points and define $\sim_{\bar{c}}$ as usual.

1. For any two tuples $\bar{i}, \bar{j} \in \mathfrak{I}^n$ and any $e \in A$, we have
$$(\triangle) \quad \bar{i} \sim_{\bar{c}} \bar{j} \Longrightarrow \phi(a_{\bar{i}}, e; b) \leftrightarrow \phi(a_{\bar{j}}, e; b).$$
2. For any convex equivalence relation $\sim$ satisfying $(\triangle)$, if $i, j \in \mathfrak{I}$ are not critical points, then $i \sim j \Longrightarrow i \sim_{\bar{c}} j$.
3. There are finitely many ϕ-critical points.
4. Let M be a model containing I, A and b. The relation $\sim_{\bar{c}}$ is Ab-definable in the pair (M, I) and each ϕ-critical point is in $\operatorname{dcl}(Ab)$ as read in that structure.

Exercise 3.38. Let $I = (a_i : i \in \mathfrak{I})$ be indiscernible, where $(\mathfrak{I}; \leq)$ is a saturated model of DLO. The group of elementary automorphisms of I naturally acts on the space of types over I. Show that the number of orbits under this action is absolutely bounded in terms of $|T|$.

References and related subjects

Honest definitions were introduced by Chernikov and Simon in [28]. Corollary 3.17 was proved first by Guingona in [46] using a different method. Proposition 3.23 (Shelah's expansion) was proved first in the o-minimal case by Baisalov and Poizat in [11] and in full generality by Shelah in [109]. Example 3.26 comes from [11]. The results on shrinking of indiscernibles were first obtained by Baldwin and Benedikt in [13], without the parameter set A, and generalized by Shelah in [109]. The proofs we give here are new.

The first application of honest definitions was to prove that certain *pairs of structures* have the NIP property. See Chernikov and Simon [28], [27]. Other results were obtained in more specific contexts by Boxall [20], Günaydin and Hieronymi [49], and Berenstein, Dolich and Onshuus [19]. See the last section of [28] for a quick overview.

In [111], Shelah shows how to *decompose an arbitrary type* over a saturated model into a finitely satisfiable part and a 'tree-like' quotient. A much more elaborate version of this decomposition appears in [110] where it is used to prove some strong theorems on NIP theories, most notably: over a saturated model, there are few types up to automorphisms, every such type is a limit of a long indiscernible sequence, and the so-called *generic pair conjecture* holds.

Chapter 4

STRONG DEPENDENCE AND DP-RANKS

We introduce the notion of dp-rank, which is similar to *weight* in stable theories. It is a measure of the complexity of a type in an NIP theory. A theory is *strongly dependent* if all types have dp-rank $< \aleph_0$. In such theories, shrinking of indiscernibles can be done at once for all formulas, using only finitely many cuts.

We assume NIP throughout this chapter.

4.1. Mutually indiscernible sequences

DEFINITION 4.1. Let $(I_t : t \in X)$ be a family of sequences and A a set of parameters. We say that the sequences $(I_t : t \in X)$ are *mutually indiscernible* over A if for each $t \in X$, the sequence I_t is indiscernible over $A \cup I_{\neq t}$.

LEMMA 4.2. *Let $(I_t : t < \alpha)$ be a family of sequences and A a set of parameters. Inductively construct I'_t realizing the EM-type of I_t over $AI'_{<t}I_{>t}$ and indiscernible over that set. Then the sequences $(I'_t : t < \alpha)$ are mutually indiscernible.*

PROOF. Fix $t < \alpha$. By construction I'_t is indiscernible over $A_t := AI_{>t}I'_{<t}$. We have to check that as we replace one by one the sequences I_s, $s > t$ by I'_s in A_t, the sequence I'_t remains indiscernible over the resulting set. If some formula with parameters in $I'_s \cup AI_{>s}I'_{<s,\neq t}$ can be used to contradict indiscernibility of I'_t, then by the EM-type assumption, such parameters can also be found in $I_s \cup AI_{>s}I'_{<s,\neq t}$; contradiction. ⊣

Example 4.3. Let $I = (a_i : i \in \mathfrak{I})$ be an A-indiscernible sequence. Let $(\mathfrak{I}_t : t \in X)$ be a family of pairwise disjoint convex subsets of $\mathfrak{I}$. For $t \in X$, let $I_t = (a_i : i \in \mathfrak{I}_t)$. Then the sequences $(I_t : t \in X)$ are mutually indiscernible over $A \cup \{a_i : i \in \mathfrak{I} \setminus \bigcup_{t \in X} \mathfrak{I}_t\}$.

Example 4.4. Let $(p_t : t < \alpha)$ be a family of global A-invariant types. Build a family $(I_t : t < \alpha)$ of sequences such that for each $t < \alpha$, I_t is a Morley sequence of p_t over $A \cup \{I_l : l < t\}$. Then the sequences $(I_t : t < \alpha)$ are mutually indiscernible over A.

[*Proof*: Observe that if I is indiscernible over A, p is a global A-invariant type and $a \models p \restriction AI$, then I is indiscernible over Aa. This is immediate by invariance of p. In particular it holds if a realizes a Morley sequence of p over AI. More generally, if the sequences $(I_t : t < \beta)$ are mutually indiscernible over A, then they remain so over Aa where $a \models p \restriction A \cup \{I_t : t < \beta\}$. Thus the result follows by induction.]

Example 4.5. Let $(p_t : t < \alpha)$ be a family of global A-invariant types such that $p_t(x) \otimes p_s(y) = p_s(y) \otimes p_t(x)$ for any $t, s < \alpha$, $t \neq s$. Build an array $(a_k^t : t < \alpha, k < \omega)$ such that $a_k^t \models p_t \restriction A \cup \{a_l^s : (l < k) \text{ or } (l = k \text{ and } s < t)\}$. For $t < \alpha$, define $I_t = (a_k^t : k < \omega)$. Note that it is a Morley sequence of p_t over A. Then the sequences $(I_t : t < \alpha)$ are mutually indiscernible over A.

[*Proof*: We show that for each $t < \alpha$, I_t is a Morley sequence of p_t over $I_{\neq t}$. Let $n < \omega$ and pick pairs

$$(k_0, t_0) > \cdots > (k_{n-1}, t_{n-1}) \in \omega \times \alpha,$$

where $\omega \times \alpha$ is ordered lexicographically. Let $c < n$ be minimal such that $t_c = t$ (we assume such a c exists). We have

$$a_{k_0}^{t_0\frown} \cdots {}^\frown a_{k_{n-1}}^{t_{n-1}} \models p_{t_0}(x_0) \otimes \cdots \otimes p_{t_{n-1}}(x_{n-1})|_A.$$

As the type p_{t_c} commutes with p_{t_m} for $m < c$, we have

$$a_{k_0}^{t_0\frown} \cdots {}^\frown a_{k_{n-1}}^{t_{n-1}} \models p_{t_c}(x_c) \otimes p_{t_0}(x_0) \otimes \cdots \otimes \widehat{p_{t_c}(x_c)} \otimes \cdots \otimes p_{t_{n-1}}(x_{n-1})|_A.$$

Where $\widehat{p_{t_c}(x_c)}$ means that this term is omitted. Therefore $a_{k_c}^{t_c} \models p_{t_c} \restriction A \cup \{a_{k_m}^{t_m} : m \neq c\}$. It follows that the sequence I_t is a Morley sequence of p_t over $A \cup \{I_s : s \neq t\}$ and we conclude as in the previous example.]

Exercise 4.6. Let $(I_t : t \in X)$ be a family of mutually indiscernible sequences. For $t \in X$, let $p_t = \lim(I_t)$. Then for $t, s \in X$, $t \neq s$, the types p_t and p_s commute (i.e., $p_t(x) \otimes p_s(y) = p_s(y) \otimes p_t(x)$).

Remark 4.7. Let $(I_t : t \in X)$ be a family of sequences, with $I_t = (a_i^t : i \in \mathfrak{I}_t)$. Assume that the sequences $(I_t : t \in X)$ are mutually indiscernible over A. Let $\mathfrak{J}$ be a linearly ordered set, and for $t \in X$, let $\sigma_t : \mathfrak{J} \to \mathfrak{I}_t$ be an increasing embedding. Then the sequence $((a_{\sigma_t(i)}^t)_{t \in X} : i \in \mathfrak{J})$ is indiscernible over A.

PROPOSITION 4.8. *Let $(I_t : t \in X)$ be a family of sequences mutually indiscernible over A. Let $b \in \mathcal{U}$ be a finite tuple. Then there is a set $X_b \subseteq X$ of size at most $|T|$ such that the sequences $(I_t : t \in X \setminus X_b)$ are mutually indiscernible over Ab.*

PROOF. For simplicity assume that the sequences I_t are sequences of finite tuples. (The case of infinite tuples can be taken care of as in the proof of Theorem 3.36). By working in M^{eq}, we may assume that they are sequences of singletons. Write $I_t = (a_i^t : i \in \mathfrak{I}_t)$. Let $\boldsymbol{P}(x)$ and $\boldsymbol{A}(x)$ be two new unary predicates, and $\boldsymbol{R}(x, y)$ a new binary predicate. Let $L' = L \cup \{\boldsymbol{P}, \boldsymbol{A}, \boldsymbol{R}\}$. We define an L'-expansion (M, P, A, R) of M by interpreting $\boldsymbol{P}(x)$ as the set $P = \{a_i^t : t \in X, i \in \mathfrak{I}_t\} \cup A$, $\boldsymbol{A}(x)$ as the set A, $\boldsymbol{R}(x, y)$ as the set $R = \{(a_i^t, a_{i'}^t) : t \in X, i < i' \in \mathfrak{I}_t\}$.

Let $(M, P, A, R) \prec (M', P', A', R')$ be some $|M|^+$-saturated elementary extension. Using the extra predicates added to the language, we see that there is some $X' \supseteq X$ and sets $\mathfrak{I}_t'$ for $t \in X'$, with $\mathfrak{I}_t \subseteq \mathfrak{I}_t'$ when $t \in X$ such that P' can be written as $P' = A' \cup \{a_i^t : t \in X', i \in \mathfrak{I}_t'\}$ and such that the sequences $(I_t' = (a_i^t : i \in \mathfrak{I}_t') : t \in X')$ are mutually indiscernible over A'.

By Corollary 3.18, there is some $P_0 \subset P'$ of size at most $|T|$ such that for any finite tuples $a, a' \in P$ we have $a \equiv_{P_0} a' \Longrightarrow a \equiv_b a'$. Let $X_b \subseteq X$ be the set of elements $t \in X$ such that some a_i^t belongs to P_0. Then X_b has size at most $|T|$ and we see that the sequences $(I_t : t \in X \setminus X_b)$ are mutually indiscernible over Ab. ⊣

Exercise 4.9. Let $(p_t : t < |T|^+)$ be a family of global A-invariant types such that $p_i(x) \otimes p_j(y) = p_j(y) \otimes p_i(x)$ for all $i \neq j$. Let q be any global invariant type. Then there is $t < |T|^+$ such that $q(x) \otimes p_t(y) = p_t(y) \otimes q(x)$.

[*Hint*: Build sequences $(I_t : t < |T|^+)$, each one being a Morley sequence of p_t over the previous ones. Then take b realizing q over those sequences. Finally realize again a family $(J_t : t < |T|^+)$ of Morley sequences of the p_t's over what has been constructed so far. Notice that the sequences $(I_t + J_t : t < |T|^+)$ are mutually indiscernible and apply Proposition 4.8.]

Exercise 4.10. Let $(I_t : t \in X)$ be a family of mutually indiscernible sequences, indexed by the same linear order $\mathfrak{I}$. Write $I_t = (a_i^t : i \in \mathfrak{I})$. Define $I = ((a_i^t : t \in X) : i \in \mathfrak{I})$. If A is any set such that I is indiscernible over A, then the sequences $(I_t : t \in X)$ are mutually indiscernible over A.

Exercise 4.11. Let $(I_t : t < \alpha)$ be a family of mutually indiscernible endless sequences. For $t < \alpha$, let $p_t = \lim(I_t)$. Let $\mathfrak{J}$ be any linear order. Construct sequences $(J_t : t < \alpha)$, $J_t = (a_i^t : i \in \mathfrak{J})$ such that for each $t < \alpha$ and $i \in \mathfrak{J}$, $a_i^t \models p_t \restriction \bigcup\{I_s : s < \alpha\} \cup \{a_j^s : s < t \text{ or } (s = t \,\&\, i < j)\}$. Then the sequences $(I_t + J_t : t < \alpha)$ are mutually indiscernible.

[*Hint*: Let $t < \alpha$ and set $A_t = \bigcup\{I_s : s < \alpha, s \neq t\} \cup \bigcup\{J_s : s < t\}$. Show that $I_t + J_t$ is indiscernible over A_t. Also, as J_t realizes some I_t-invariant type over A_t, any sequence in A_t indiscernible over I_t remains indiscernible over $I_t + J_t$. The result follows by induction on $t < \alpha$.]

4.2. Dp-ranks

DEFINITION 4.12. Let p be a partial type over a set A, and let κ be a (finite or infinite) cardinal. We say $\text{dp-rk}(p, A) < \kappa$ if for every family $(I_t : t < \kappa)$ of mutually indiscernible sequences over A and $b \models p$, there is $t < \kappa$ such that I_t is indiscernible over Ab.

If $b \in \mathcal{U}$, then dp-rk(b/A) stands for dp-rk$(\mathrm{tp}(b/A), A)$.

We implicitly allow the sequences I_t to be sequences of infinite tuples. This does not actually make any difference, because if an indiscernible sequence $I = (\bar{a}_i : i \in \mathfrak{I})$ is not indiscernible over some b, then there are finite $a'_i \subseteq \bar{a}_i$ such that the sequence $(a'_i : i \in \mathfrak{I})$ is indiscernible, but not indiscernible over b.

OBSERVATION 4.13. *(Not assuming NIP.) The following are equivalent:*

- *the theory T is NIP;*
- *for every finitary type p and set A, we have* dp-rk$(p, A) < |T|^+$;
- *for every type p and A, there is some κ such that* dp-rk$(p, A) < \kappa$.

PROOF. The implication from the first dot to the second follows from Proposition 4.8. The implication from the second dot to the third is immediate.

For the last implication, assume that the formula $\phi(x; y)$ has IP. Let κ be any cardinal and $\mathfrak{I} = \omega \times \kappa$ ordered lexicographically. Then we can find a $|y|$-tuple b and an indiscernible sequence $(a_i : i \in \mathfrak{I})$ such that $\phi(a_i; b)$ holds if and only if i is of the form $(0, \alpha)$, $\alpha < \kappa$. The sequences $I_t = (a_{(n,t)} : n < \omega)$ are mutually indiscernible, but none stays indiscernible over b. Hence dp-rk$(p, \emptyset) \geq \kappa$. ⊣

We say that dp-rk$(p, A) = \kappa$ if dp-rk$(p, A) < \kappa^+$, but not dp-rk$(p, A) < \kappa$. Note that it may happen that we have dp-rk$(p, A) = \kappa$ for no value of κ. For example, assume that for every integer n we can find a family $(I_t : t < n)$ of mutually indiscernible sequences over A, none of which is indiscernible over Ab, but we cannot find such a family of size $\aleph_0$. Then we would have dp-rk$(b/A) < \aleph_0$, but dp-rk$(b/A) \geq n$ for all n. One could probably write "dp-rk$(b/A) = \aleph_0^-$" in this situation (as done for example in [2]), but we will not use that notation. See [65] for a concrete example of this behavior.

LEMMA 4.14. *Let p be a partial type over A and κ any cardinal. Let also $A \subseteq B$. Then* dp-rk$(p, A) < \kappa \iff$ dp-rk$(p, B) < \kappa$.

PROOF. Assume that dp-rk$(p, A) < \kappa$. Let $(I_t : t < \kappa)$ be mutually indiscernible over B. Let $\bar{b}$ enumerate B and let $d \models p$. For $t < \kappa$, write $I_t = (a^t_i : i \in \mathfrak{I}_t)$ and define the sequence $J_t = (a^t_i{}^\frown \bar{b} : i \in \mathfrak{I}_t)$. Then the sequences $(J_t : t < \kappa)$ are mutually indiscernible over A. By hypothesis, there is $t < \kappa$ such that J_t is indiscernible over Ad. This implies that I_t is indiscernible over Bd.

Conversely, assume that we have a witness of dp-rk$(p, A) \geq \kappa$. Namely, we have some $d \models p$ and a family $(I_t : t < \kappa)$ of sequences, mutually indiscernible over A such that no I_t is indiscernible over Ad. Using Lemma 4.2 over the base B, we can construct sequences $(I'_t : t < \kappa)$ mutually indiscernible over B such that $\mathrm{tp}((I'_t)_{t<\kappa}/A) = \mathrm{tp}((I_t)_{t<\kappa}/A)$. Then any d' such that $\mathrm{tp}(d'(I'_t)_{t<k}/A) = \mathrm{tp}(d(I_t)_{t<k}/A)$ witnesses that dp-rk$(p, B) \geq \kappa$. ⊣

Remark 4.15. If $p \subseteq q$ are partial types over B, then dp-rk$(p, B) < \kappa$ implies dp-rk$(q, B) < \kappa$. It follows that if $A \subseteq B$ and $b \in \mathcal{U}$, then dp-rk$(b/A) < \kappa$ implies dp-rk$(b/B) < \kappa$.

Example 4.16. In the theory T_{dt} of dense trees, the formula $x = x$ has dp-rank 1, as can easily be checked by inspection.

Let T be the model-companion of the theory of two linear orders in the language $L = \{\leq_1, \leq_2\}$. Then $x = x$ has dp-rank 2 in T and in fact no type in T has dp-rank exactly 1.

PROPOSITION 4.17. *Let p be a partial type over A and let κ be any cardinal. Then we have* dp-rk$(p, A) < \kappa$ *if and only if for any family $(I_t : t \in X)$ of sequences, mutually indiscernible over A and any $b \models p$, there is $X_0 \subseteq X$ of size $< \kappa$ such that $(I_t : t \in X \setminus X_0)$ are mutually indiscernible over Ab.*

PROOF. It is obvious that the property considered implies dp-rk$(p, A) < \kappa$. We show the converse.

Case 1: κ is infinite.

Assume that dp-rk$(p, A) < \kappa$ and assume that $(I_t : t \in X)$ and b give a counterexample to what we have to prove. Without loss, $X = \kappa$. We can build an increasing sequence $(\delta_t : t < \kappa)$ of ordinals, and a sequence $(\Delta_t : t < \kappa)$ of finite subsets of κ such that:

- for all $t < \kappa$, the sequence I_{δ_t} is not indiscernible over $\{b\} \cup \bigcup\{I_s : s \in \Delta_t\}$;
- for all $t < t' < \kappa$, $(\Delta_t \cup \{\delta_t\}) \cap (\Delta_{t'} \cup \{\delta_{t'}\}) = \emptyset$.

Let $B = A \cup \bigcup\{I_s : s \in \Delta_t, t < \kappa\}$. Then the sequences $(I_{\delta_t} : t < \kappa)$ are mutually indiscernible over B, and for each $t < \kappa$, I_{δ_t} is not indiscernible over Bb. This contradicts dp-rk$(p, B) =$ dp-rk$(p, A) < \kappa$.

Case 2: $\kappa = n + 1$ is finite.

Assume that dp-rk$(p, A) < n + 1$ and let $(I_t : t \in X)$ be mutually indiscernible over A. We show the result by induction on $|X|$.

First we deal with the finite case. If $|X| \leq n$, the result is obvious as we may take $X_0 = X$.

Assume that $|X| = n + k + 1$, and we have shown the result for sets of cardinality $\leq n+k$. Let $(I_t : t < n+k+1)$ be mutually indiscernible over A and let $b \models p$. We may assume that the sequences I_t are endless. For $t < n + k + 1$, let p_t denote the limit type $\lim(I_t)$. Construct sequences $I'_t = (c^t_k : k < \omega)$, such that $c^t_k \models p_t \upharpoonright Ab \cup \{I_t : t < n + k + 1\} \cup \{I'_s : s < t\} \cup \{c^t_l : l > k\}$ (so those are Morley sequences of each p_t, but read backwards).

Then the sequences $(I_t + I'_t : t < n + k + 1)$ are mutually indiscernible over A (see Exercise 4.11).

Let $B = A \cup \bigcup\{I'_t : t < n + k + 1\}$.

As dp-rk$(p, B) =$ dp-rk$(p, A) \leq n$, there is some $t < n + k + 1$ such that I_t is indiscernible over Bb. Without loss, assume that $t = 0$.

By induction hypothesis, working over the base set AI_0, there is some $X_0 \subset \{1, \ldots, n+k\}$ of size at most n such that the sequences $(I_t + I'_t : t \in \{1, \ldots, n+k\} \setminus X_0)$ are mutually indiscernible over $AI_0 b$. Again, without loss, $X_0 \subseteq \{k+1, \ldots, k+n\}$ so that the sequences $(I_t + I'_t : 0 < t < k+1)$ are mutually indiscernible over $AI_0 b$. If I_0 is indiscernible over $Ab \cup \bigcup\{I_t : 0 < t < k+1\}$ then the sequences $(I_t : t < k+1)$ are mutually indiscernible over Ab and we are done.

Otherwise, there is some formula $\phi(x_1, \ldots, x_n, \bar{d}) \in L(Ab \cup \bigcup\{I_t : 0 < t < k+1\})$ which witnesses that I_0 is not indiscernible over $Ab \cup \bigcup\{I_t : 0 < t < k+1\}$. As the sequences $(I_t + I'_t : 0 < t < k+1)$ are mutually indiscernible over $AI_0 b$, we can move the parameters in ϕ belonging to some I_k to parameters in I'_k. Formally, there is some $\bar{d}' \equiv_{AI_0 b} \bar{d}$ such that $\bar{d}' \in Ab \cup \bigcup\{I'_t : 0 < t < k+1\}$. Then $\phi(x_1, \ldots, x_n, \bar{d}')$ witnesses the fact that I_0 is not indiscernible over Bb. This contradicts the assumption on I_0.

Now assume that $|X| = \lambda$ is infinite. Write X as an increasing union $X = \bigcup_{i<\alpha} X^i$ where each X^i has size $< \lambda$. By induction, for every $i < \alpha$, there is a subset X_0^i of X_i of size $\leq n$ such that the sequences in $(I_t : t \in X^i \setminus X_0^i)$ are mutually indiscernible over Ab. If the sequences $(I_t : t \in X)$ are not mutually indiscernible over Ab, there is some finite $\Delta_0 \subset X$ such that already $(I_t : t \in \Delta_0)$ are not mutually indiscernible over Ab. Hence every X_0^i must intersect Δ_0 and there is some $t_0 \in \Delta_0$ contained in cofinally many of the X_0^i's. If the sequences $(I_t : t \in X \setminus \{t_0\})$ are mutually indiscernible over Ab, we are done. Otherwise, we again find some finite $\Delta_1 \subset X \setminus \{t_0\}$ witnessing it. There is some $t_1 \in \Delta_1$ such that $\{t_0, t_1\}$ is included in cofinally many X_0^i's. Iterating, we obtain $t_0, \ldots, t_{n-1}$ which are in cofinally many X_0^i's and one checks at once that the sequences $(I_t : t \in X \setminus \{t_0, \ldots, t_{n-1}\})$ are mutually indiscernible over Ab. ⊣

We recall that the notation compl($\mathfrak{I}$) is used to denote the completion of the linear order $\mathfrak{I}$. Recall also the definition of the equivalence relation $\sim_{\bar{c}}$ as defined before Proposition 3.32. In the statement and proof of the next theorem, it will be convenient to also consider $\sim_{\bar{c}}$ as an equivalence relation on $I = (a_i : i \in \mathfrak{I})$ in the obvious way: $a_i \sim_{\bar{c}} a_j$ if $i \sim_{\bar{c}} j$.

THEOREM 4.18. *Let p be a partial type over A and κ a cardinal (possibly finite). The following are equivalent:*

(i) dp-rk$(p, A) < \kappa$.

(ii)$_0$ *If $(I_t : t \in X)$ are mutually indiscernible over A and $b \models p$, then there is some $X_0 \subseteq X$ of size $< \kappa$ such that for $t \in X \setminus X_0$, all the members of I_t have the same type over Ab.*

(ii)$_1$ *Same as above, but change the conclusion to: for $t \in X \setminus X_0$, the sequence I_t is indiscernible over Ab.*

(ii)$_2$ *Same as above, but change the conclusion to: the sequences $(I_t : t \in X \setminus X_0)$ are mutually indiscernible over Ab.*

(iii)$_0$ *If $I = (a_i : i \in \mathfrak{I})$ is an A-indiscernible sequence (of possibly infinite tuples), there is some $\bar{c} \in \operatorname{compl}(\mathfrak{I})$, $|\bar{c}| < \kappa$ such that $i \sim_{\bar{c}} j \Rightarrow \operatorname{tp}(a_i/Ab) = \operatorname{tp}(a_j/Ab)$.*

(iii)$_1$ *Same as above, with the conclusion that each $\sim_{\bar{c}}$-class of I is indiscernible over Ab.*

(iii)$_2$ *Same as above, with the conclusion that the $\sim_{\bar{c}}$-classes of I which are infinite are mutually indiscernible over Ab.*

PROOF. (i) $\Rightarrow$ (ii)$_2$: This is exactly Proposition 4.17.

(ii)$_2 \Rightarrow$ (ii)$_1 \Rightarrow$ (ii)$_0$: Clear.

(ii)$_0 \Rightarrow$ (i): Assume (ii)$_0$ and assume for a contradiction that there is some family $(I_t : t < \kappa)$ of sequences, mutually indiscernible over A and $b \models p$ such that for each t, I_t is not indiscernible over Ab. Without loss, each I_t is indexed by $\mathbb{Q}$: $I_t = (a_i^t : i \in \mathbb{Q})$. Then for each t, there are $\alpha_t < \beta_t \in \mathbb{Q}$ such that $\operatorname{tp}(a_{\alpha_t}/AC_tb) \neq \operatorname{tp}(a_{\beta_t}/AC_tb)$ where $C_t = \{a_i^t : i < \alpha_t \text{ or } i > \beta_t\}$. Set $I'_t = (a_i^{t\frown}C_t : \alpha_t \leq i \leq \beta_t)$. Then the family $(I'_t : t < \kappa)$ contradicts (ii)$_0$.

(iii)$_2 \Rightarrow$ (iii)$_1 \Rightarrow$ (iii)$_0$: Clear.

(ii)$_2 \Rightarrow$ (iii)$_2$: Assume (ii)$_2$ and let $I = (a_i : i \in \mathfrak{I})$ be indiscernible over A, and $b \models p$. Without loss, $\mathfrak{I}$ is a dense complete order, so $\operatorname{compl}(\mathfrak{I}) = \mathfrak{I}$.

For every $i \in \mathfrak{I}$, let p_i^- be the limit type of the sequence $(a_j : j \in \mathfrak{I}, j < i)$. Similarly, let p_i^+ be the limit type of the sequence $(a_j : j \in \mathfrak{I}, j > i)$ read in the reverse order.

Step 1: We blow up each point of I to an indiscernible sequence: For each $i \in \mathfrak{I}$, set $d_0^i = a_i$. Let $(d_n^i : 0 < n < \omega)$ be a Morley sequence of q_i^+ over everything constructed so far and $(d_{-n}^i : 0 < n < \omega)$ be a Morley sequence of q_i^- over everything. Set $A_i = (d_n^i : n \in \mathbb{Z})$. It is an indiscernible sequence and so is the concatenation $\sum_{i \in \mathfrak{I}} A_i$.

Step 2: We apply (ii)$_2$: The sequences $(A_i : i \in \mathfrak{I})$ are mutually indiscernible, hence by (ii)$_2$ there is some $\bar{c} \subseteq \mathfrak{I}$ of size $< \kappa$ such that the sequences $(A_i : i \in \mathfrak{I} \setminus \bar{c})$ are mutually indiscernible over Ab.

Step 3: We show that the family of $\sim_{\bar{c}}$ classes of I which are not reduced to a point are mutually indiscernible. Assume this is not the case. Then there is some formula $\phi(x) \in L(AIb)$, indices $k < l \in \mathfrak{I}$ such that:

- $a_k \sim_{\bar{c}} a_l$;
- $\models \phi(a_k) \wedge \neg\phi(a_l)$;
- all the parameters in ϕ come from $Ab \cup \{a_i : i < k\} \cup \{a_i : i > l\}$ and are disjoint from $\bar{c}$.

Let $i_* = \sup\{i < l : \models \phi(a_i)\}$ and $\varepsilon \in \{0, 1\}$ such that $a_{i_*} \models \phi(x)^\varepsilon$. Then $k \leq i_* \leq l$ and as $k \sim_{\bar{c}} l$, i_* does not lie in $\bar{c}$. We look to the left and to the right of a_{i_*}. Both limit types $p_{i_*}^-$ and $p_{i_*}^+$ must satisfy $\phi(x)^\varepsilon$ since the sequence A_{i_*} is indiscernible over the parameters of ϕ. Moreover, by definition of i_*, if

$k < i_*$, then $p^-_{i_*} \vdash \phi(x)$ and if $i_* < l$, then $p^+_{i_*} \vdash \neg\phi(x)$. In all cases, we obtain a contradiction.

(iii)$_0 \Rightarrow$ (ii)$_0$: Let $(I_t : t < \alpha)$ be mutually indiscernible over A, and let $b \models p$. Write $I_t = (a^t_i : i \in \mathfrak{I}_t)$, where the indexing orders $\mathfrak{I}_t$ are disjoint. Let $\mathfrak{J}$ denote the linear order $\sum_{t<\alpha} \mathfrak{I}_t$. We may expand the family $(I_t : t < \alpha)$ into a family $(J_t : t < \alpha)$ of sequences, mutually indiscernible over A such that $J_t = (a^t_i : i \in \mathfrak{J})$ (in particular, I_t is a subsequence of J_t). Let $J = (a^{0\frown}_i a^{1\frown}_i \cdots : i \in \mathfrak{J})$. Then J is an indiscernible sequence. By assumption (iii)$_0$, there is some $\bar{c} \subseteq \text{compl}(\mathfrak{J})$ of size $< \kappa$ such that

$$i \sim_{\bar{c}} j \Longrightarrow \text{tp}(a^{0\frown}_i a^{1\frown}_i \cdots /Ab) = \text{tp}(a^{0\frown}_j a^{1\frown}_j \cdots /Ab).$$

Let $X_0 \subset \alpha$ be the set of $t < \alpha$ such that $\bar{c} \cap \mathfrak{I}_t \neq \emptyset$. Then $|X_0| < \kappa$ and $\bar{c}$ satisfies the requirement of (ii)$_0$. ⊣

Remark 4.19. In (iii)$_2$, we insist that the *infinite* classes are mutually indiscernible. If the indiscernible sequence I is densely ordered, then all classes are either infinite or reduced to a point. So the conclusion says nothing about the latter ones. It may happen that $\bar{c}$ can be chosen so that the infinite classes are mutually indiscernible over (A and) the finite ones. This is a stronger property, which we could call dp-rk$^+(p, A) < \kappa$ (because it is linked with what Shelah calls strongly$^+$-dependent). Here is an example where this stronger property does not hold.

Let T be ACVF, the theory of algebraically closed valued fields (see Section A.2). Let $(a_i : i \in \mathbb{Q})$ be an indiscernible sequence of elements such that $v(a_i) < v(a_j)$ for all $i < j$. Then we can find a point b such that $v(b - (a_0 + \cdots + a_n)) > v(a_n)$ for all $n < \omega$. (We then necessarily have $v(b - (a_0 + \cdots + a_n)) = v(a_{n+1})$. In other words, b is a pseudo-limit of the sequence $(a_0 + \cdots + a_n : n < \omega)$).

The point b splits the sequence I into three pieces: $(a_i : i < 0)$, (a_0) and $(a_i : i > 0)$. The two infinite ones are mutually indiscernible over b. However, the sequence $(a_i : i > 0)$ is not indiscernible over $a_0 b$. In fact that tuple breaks it again into three pieces: $(a_i : 0 < i < 1)$, (a_1) and $(a_i : i > 1)$. And we can go on.

This shows that ACVF is not strongly$^+$-dependent in Shelah's terminology. However it is strongly dependent, even dp-minimal (see definitions below).

PROPOSITION 4.20. *Let $a, b \in \mathcal{U}$, let A be a small set of parameters and κ_1, κ_2 be two cardinals such that* dp-rk$(b/A) < \kappa_1$ *and* dp-rk$(a/Ab) < \kappa_2$. *Then* dp-rk$(a, b/A) < \kappa_1 + \kappa_2 - 1$.

PROOF. Here $\kappa - 1$ is equal to κ when κ is infinite.

We use condition (ii)$_2$. Let $(I_t : t \in X)$ be mutually indiscernible over A. There is some $X_0 \subseteq X$ of size $< \kappa_1$ such that the sequences I_t, $t \in X \setminus X_0$ are indiscernible over Ab. Then there is some $X_1 \subseteq X \setminus X_0$ of size $< \kappa_2$ such that the sequences I_t, $t \in X \setminus (X_0 \cup X_1)$ are indiscernible over Aab. ⊣

Note in particular that the hypothesis of the proposition are satisfied if dp-rk$(a/A) < \kappa_1$ and dp-rk$(b/A) < \kappa_2$.

Another characterization of dp-ranks is by Shelah-style arrays.

DEFINITION 4.21. Let $p(y)$ be a partial type over a set A. We define $\kappa_{\text{ict}}(p, A)$ as the minimal κ such that the following does not exist:

- formulas $\phi_\alpha(x_\alpha; y)$;
- an array $(a_i^\alpha : i < \omega, \alpha < \kappa)$ of tuples, with $|a_i^\alpha| = |x_\alpha|$;
- for every $\eta : \kappa \to \omega$, a tuple $b_\eta \models p$ such that we have

$$\models \phi_\alpha(a_i^\alpha; b_\eta) \iff \eta(\alpha) = i.$$

(Such a family of tuples and formulas will be called an *ict-pattern* for p. The cardinal κ is the *length* of the pattern).

PROPOSITION 4.22. *For any partial type p over A and cardinal κ, we have* dp-rk$(p, A) < \kappa$ *if and only if* $\kappa_{\text{ict}}(p, A) \leq \kappa$.

PROOF. Assume there is an ict-pattern for p of length κ. By expanding and extracting, we may assume that the rows $(a_i^\alpha : i < \omega)$, $\alpha < \kappa$, are mutually indiscernible over A. Then for any η, b_η witnesses that dp-rk$(p, A) \geq \kappa$.

Conversely, assume that dp-rk$(p, A) \geq \kappa$. By condition (ii)$_0$ of Theorem 4.18 we can find:

- sequences $(I_t : t < \kappa)$, mutually indiscernible over A, without loss $I_t = (a_i^t : i < \omega)$ are sequences of finite tuples;
- formulas $\phi_t(x_t; y)$;
- a tuple $b \models p$ such that $\models \phi_t(a_0^t; b) \wedge \neg\phi_t(a_1^t; b)$.

Let $\psi_t(x_t, x_t'; y) = \phi_t(x_t; y) \wedge \neg\phi_t(x_t'; y)$. For $t < \kappa$, let $J_t = (a_{2i}^t{}^\frown a_{2i+1}^t : i < \omega)$. By NIP, at most finitely many tuples $a^\frown a'$ in J_t satisfy $\psi_t(x_t, x_t'; b)$. Removing some if necessary, we may assume that $a_0^t{}^\frown a_1^t$ is the only element of J_t for which that formula holds. By mutual indiscernibility of the sequences $(J_t : t < \kappa)$, for every $\eta : \kappa \to \omega$, we can find some $b_\eta \models p$ such that, for all $t < \kappa$,

$$\models \psi_t(a_{2i}^t, a_{2i+1}^t; b_\eta) \iff \eta(t) = i.$$

This shows that $\kappa_{\text{ict}}(p, A) > \kappa$. ⊣

The difference between dp-rk and κ_{ict} is only due to the convention chosen for the definition of dp-rank. One argument in favor of this definition, as opposed to that of κ_{ict}, is that we want $x = x$ in the theory of equality to have rank 1 and not 2.

4.3. Strongly dependent theories

DEFINITION 4.23. The NIP theory T is *strongly dependent* if for any finite tuple of variables x, we have dp-rk$(x = x, \emptyset) < \aleph_0$.

Example 4.24. Let $L = \{E_i : i < \omega\}$ where the E_i's are binary predicates. Let T be the L-theory stating that each E_i defines an equivalence relation with infinitely many classes, each of which is infinite. Consider $T_1 \supset T$ stating that $(\forall xy) x E_{i+1} y \rightarrow x E_i y$, and each E_i-class is split into infinitely many E_{i+1} classes.

Consider also $T_2 \supset T$ stating that the E_i's are cross-cutting, that is: given $a_0, \dots, a_{n-1}$, there is a such that $a E_k a_k$ for every $k < n$.

Then T_1 is strongly dependent, but not T_2 (and note that both are stable).

Remark 4.25. If T is a superstable theory, then it is strongly dependent, but the converse does not hold, even assuming stability, as witnessed by T_1 in the previous example. See Section 4.4.

PROPOSITION 4.26. *If* dp-rk$(x = x, \emptyset) < \aleph_0$ *for every variable* x *with* $|x| = 1$, *then* T *is strongly dependent.*

PROOF. This follows immediately from Proposition 4.20. ⊣

An extreme case of strongly dependent theories are dp-minimal theories.

DEFINITION 4.27. The theory T is *dp-minimal* if dp-rk$(x = x, \emptyset) = 1$, for x a singleton.

Example 4.28. The following theories are dp-minimal (some proofs are given in Appendix A. See also [35]):

- any o-minimal theory;
- the theory T_{dt} of dense trees;
- any C-minimal theory, in particular ACVF, the theory of algebraically closed valued fields;
- the theory $Th(\mathbb{Q}_p)$ of the p-adics;
- the theory of $(\mathbb{Z}, 0, 1, +, \leq)$.

We point out some link with honest definitions. First an easy lemma.

LEMMA 4.29. *Let* $A \subset \mathcal{U}$. *Let* $(p_i : i < \omega)$ *be a family of* A*-invariant types and let* $(c_i, d_i : i < \omega)$ *and* $b \in \mathcal{U}$ *be finite tuples such that:*

- $c_i \models p_i \restriction Ac_{<i}d_{<i}$;
- $d_i \models p_i \restriction Ac_{\leq i}d_{<i}$;
- $\text{tp}(c_i/Ab) \neq \text{tp}(d_i/Ab)$.

Then T *is not strongly dependent.*

PROOF. Build sequences $I_i = (e^i_j : j < \omega)$ for $i < \omega$ such that I_i is a Morley sequence of p_i over $AI_{<i}$. Then we see that $\text{tp}((c_i, d_i : i < \omega)/A) = \text{tp}((e^i_0, e^i_1 : i < \omega)/A)$. Therefore, composing by an automorphism, we may assume that $e^i_0 = c_i$ and $e^i_1 = d_i$ for all $i < \omega$. Then the sequences $(I_i : i < \omega)$ are mutually indiscernible over A, but none of them remains indiscernible over Ab. This implies that T is not strongly dependent. ⊣

PROPOSITION 4.30. *Let T be strongly dependent. Let $M \models T$, $A \subseteq M$, $b \in M$ a finite tuple, and $(M, A) \prec (M', A')$ some $|M|^+$-saturated extension. Then there is $A_0 \subseteq A'$ finite such that every formula $\phi(x; b) \in L(b)$ has an honest definition over A with parameters in AA_0.*

The conclusion means that we can find $\psi(x; c) \in L(AA_0)$ such that $\phi(A; b) \subseteq \psi(A'; c) \subseteq \phi(A'; b)$.

PROOF. For x a finite tuple of variables, let $S^A_x \subset S_x(\mathcal{U})$ be the set of global types, finitely satisfiable in A. By Lemma 4.29, we can build a maximal sequence $(c_i, d_i : i < N)$ such that there are types $p_i \in S^A_{x_i}$ with:

- $c_i \models p_i \restriction Ac_{<i}d_{<i}$;
- $d_i \models p_i \restriction Ac_{\leq i}d_{<i}$;
- $\mathrm{tp}(c_i/Ab) \neq \mathrm{tp}(d_i/Ab)$.

Set $A_0 = \{c_i, d_i : i < N\}$.

Let now $\phi(x; b)$ be given. Then for any type $q \in S^A_x$, there is a truth value $\varepsilon_q \in \{0, 1\}$ such that for any realization a of $q|_{AA_0}$ in A', we have $\models \phi(a; b)^{\varepsilon_q}$. By saturation of the pair (M', A'), there is a formula $\psi_q(x) \in q|_{AA_0}$ such that $(M', A') \models (\forall x \in \boldsymbol{P})\psi_q(x) \to \phi(x; b)^{\varepsilon_q}$.

We then conclude exactly as in the proof of Theorem 3.13. (Namely, we extract from $\{\psi_q(x) : \varepsilon_q = 1\}$ a finite subcover, and take the union of those formulas.) ⊣

Note that the converse does not hold since any stable theory satisfies the conclusion of the proposition, but not all stable theories are strongly dependent.

We end this section with an application to definable groups.

We say that a group G is of finite exponent if there exists $n < \omega$ such that $g^n = e$ for all $g \in G$.

PROPOSITION 4.31. *Let G be a dp-minimal group. Then there is a definable, abelian, normal subgroup H such that the quotient G/H is of finite exponent.*

PROOF.

CLAIM. *For any two definable subgroups H and K of G, one of $[H : H \cap K]$ or $[K : H \cap K]$ is finite.*

PROOF OF CLAIM. Assume not. Then we can find two sequences $(a_i : i < \omega)$ of points of K and $(b_i : i < \omega)$ of points of H such that the cosets $\{a_i H : i < \omega\}$ are distinct, as well as the cosets $\{Kb_i : i < \omega\}$. Note that for any $i, j < \omega$, there is a point in $a_i H \cap Kb_j$, namely $a_i b_j$. As two distinct cosets of the same group are disjoint, we have $a_i b_j \in a_{i'} H \cap Kb_{j'} \iff (i, j) = (i', j')$ and we have obtained an ict-pattern of length 2. By Proposition 4.22, this shows that $\text{dp-rk}(x = x, \emptyset) \geq 2$, contradicting dp-minimality. ⊣ (Claim)

It follows that if H and K are two definable subgroups, $a \in H$ and $b \in K$, then there is n such that either $a^n \in K$ or $b^n \in H$. If $a \in G$, we let

$C(a) = \{g \in G : ga = ag\}$. Applying the previous observation to the uniform family $\{C(a) : a \in G\}$ and compactness yields: there is some integer n such that for any two elements $a, b \in G$, there is $k \leq n$ such that either $a^k \in C(b)$ or $b^k \in C(a)$. In particular, if $N = n!$, then a^N and b^N commute.

Let $H = C(C(G^N))$. Note that $C(X)$ is normal whenever X is, therefore H is normal. One easily checks that it is abelian, and the quotient G/H has finite exponent bounded by N. ⊣

4.4. Superstable theories

In this section, we say a few words about superstability and show the difference with strong dependence. The only proof here assumes some familiarity with stability theory and forking. No part of this is used elsewhere in the book. For more information, see any text on stability theory, e.g. [95].

DEFINITION 4.32. A theory T is *superstable* if it is stable and if for every type $p \in S(A)$ in finitely many variables there is some finite $A_0 \subseteq A$ such that p does not fork over A_0.

FACT 4.33. *A stable theory T is superstable if and only if there is no sequence of finitary types $(p_t : t < \omega)$ such that p_{t+1} extends p_t and forks over the domain of p_t.*

LEMMA 4.34. (*Not assuming NIP.*) *A theory T is superstable if and only if one cannot find a finite tuple b and a family $(I_t : t < \omega)$ such that:*

- *for each $t < \omega$, the sequence I_t is indiscernible over $\{I_s : s < t\}$;*
- *for each $t < \omega$, the sequence I_t is not indiscernible over b.*

PROOF. Assume that T is stable and one can find such b and $(I_t : t < \omega)$. For $t < \omega$, let $p_t = \text{tp}(b/I_{\leq t})$. As the sequence I_t is indiscernible over $I_{<t}$, the type p_t forks over $I_{<t}$. Then by Fact 4.33, T is not superstable.

Conversely assume that one cannot find such b and $(I_t : t < \omega)$. First we deduce that T is stable. Assume not, then we can find a finite tuple b, an indiscernible sequence $(a_i : i \in \mathbb{R})$ and a formula $\phi(x; y)$ such that $\phi(b; a_i)$ holds if and only if $i < 0$. But then we contradict the hypothesis by setting

$$I_t = \left(a_i : i \in \left(-\frac{1}{t}, -\frac{1}{t+1}\right) \cup \left(\frac{1}{t+1}, \frac{1}{t}\right)\right).$$

If T is not superstable, we can find an increasing sequence $(A_t : t < \omega)$ of sets and a sequence $(p_t : t < \omega)$ of types such that p_t is over A_t, p_{t+1} extends p_t and forks over A_t. Let $b \models \bigcup_{t<\omega} p_t$. We build a sequence $(I_t : t < \omega)$ of indiscernible sequences as follows. Assume we have built $(I_t : t < s)$. Set $A'_{s-1} = A_{s-1} \cup \bigcup\{I_t : t < s\}$. The construction will ensure that for $t \geq s$, p_t forks over A'_{s-1}. As forking is equal to dividing in stable theories, there is a formula $\phi(x;a) \in p_s$ which divides over A'_{s-1}. Let $I_s = (a_i : i < \omega)$ be an A'_{s-1}-indiscernible sequence witnessing dividing with $a_0 = a$. Choose I_s such that $I_s \mathop{\smile\!\!\!\!\!\!\,|}_{A'_s} A_{<\omega}$. Then by properties of forking, for $t > s$, the type p_t forks over $A'_s \cup I_s$, and we can iterate. At the end, we contradict the hypothesis. $\dashv$

COROLLARY 4.35. *If T is superstable, then it is strongly dependent.*

Example 4.36. The theory T_1 presented in Example 4.24 (refining equivalence relations) is strongly dependent, stable, but not superstable.

The relation between dp-rank and weight in the stable context is explained in [86].

4.5. Exercises

Exercise 4.37. Let T be a one-sorted complete NIP theory which admits elimination of quantifiers in some language L and such that $\mathrm{acl}(A) = A$ for any $A \subset \mathcal{U}$. Assume that the binary relation symbol $\leq$ does not appear in L and set $L' = L \cup \{\leq\}$. Let T' be the L'-theory consisting of T along with axioms saying that $\leq$ defines a dense linear order without endpoints and every L-definable infinite set is $\leq$-dense.

0. Show that T' is complete, NIP, and admits quantifier elimination in L'.
1. Show that $\text{dp-rk}_{T'}(x = x) \geq \text{dp-rk}_T(x = x)$.
2. Show that if equality holds, then T is stable and dp-minimal.

Exercise 4.38 (dp-rank and acl-dimension). We say that acl satisfies exchange if for any two singletons a, b in $\mathcal{U}$ and any $A \subset \mathcal{U}$, $b \in \mathrm{acl}(Aa) \setminus \mathrm{acl}(A)$ implies $b \in \mathrm{acl}(Ab) \setminus \mathrm{acl}(A)$. When this holds, we define the acl-dimension of a set A as the smallest cardinality of an $A_0 \subseteq A$ such that $A \subseteq \mathrm{acl}(A_0)$.

We say that dp-rank is additive if the equality

$$\text{dp-rk}(\bar{a}, \bar{b}/A) = \text{dp-rk}(\bar{a}/A) + \text{dp-rk}(\bar{b}/A\bar{a})$$

holds for all $\bar{a}, \bar{b}, A$.

Let T be a dp-minimal theory. Show that dp-rank is additive if and only if acl satisfies exchange and that in this case dp-rank and acl-dimension coincide.

Exercise 4.39 (Non-continuity of dp-rank). Give an example of a theory T and a type $p(x)$ in it such that $\text{dp-rk}(p) = 1$, but $\text{dp-rk}(\phi(x)) = 2$ for any formula in p.

(One can find T of dp-rank 2. It can be shown that this is the best possible: no such type p or formula $\phi(x;y)$ can be found in a dp-minimal theory.)

References and related subjects

Strong dependence was defined by Shelah in [109] as a tentative analog of superstable, and further studied in [102]. In that paper, he defines a whole family of "dp-ranks" which depend not only on a type p and a set I, but also on a set B of indiscernible sequences, or pairs of points, "split" by p. This allows to him make constructions where the rank drops. He uses them to prove that in a strongly dependent theory, any long enough sequence contains an infinite indiscernible subsequence. The notion of dp-rank we present here is a simplification of those ranks, it is also similar to Shelah's κ_{ict}, but restricted to a type. It was first explicitly defined by Usvyatsov in [116] and it coincides in NIP theories with what Adler had previously called *burden* in [2].

Proposition 4.17 and the additivity of dp-rank that follows from it is from Kaplan, Onshuus and Usvyatsov [65]. (In which the convention for dp-rk(p) is slightly different. To be precise, dp-rk$(p) < \kappa$ for us is equivalent to dp-rk$(p) \leq \kappa - 1$ in the sense of [65]. In particular, for infinite cardinals, that latter definition agrees with κ_{ict}, but not for finite ones.) The equivalent characterizations given in Theorem 4.18 are from Shelah [109], except for $(\text{ii})_2$ and $(\text{iii})_2$ which make use of Proposition 4.17.

The notion of a dp-minimal theory originates in Shelah's work, but was precisely defined and studied by Onshuus and Usvyatsov in [86].

Proposition 4.30 is new. Proposition 4.31 is from Simon [112].

Shelah investigates in [102] a variety of *other notions of strong dependence*, in particular one which is called strongly$^+$-dependent (and was alluded to in Remark 4.19). He conjectures that strongly$^+$-dependent fields are algebraically closed and gives a conjectural classification of strongly dependent fields (algebraically closed, real closed or admitting a Henselian valuation).

Shelah proves in [102] that if a theory is strongly dependent, then from any long enough sequence, one can *extract an indiscernible subsequence*. Counterexamples in the general NIP case are given by Kaplan and Shelah in [68] and [67].

Basic facts and examples about *dp-minimal theories* can be found in Dolich, Lippel and Goodrick's paper [35]. Concerning dp-minimal ordered theories, Goodrick proves in [45] a monotonicity theorem on definable functions. Further results about dp-minimal groups appear in Simon [112].

A stronger notion than dp-minimality is *VC-minimality* which was introduced by Adler in [3]. It is studied by Guingona and Laskowski in [48], where in particular reducts of VC-minimal theories are characterized as being *convexely orderable*. Further work by Flenner and Guingona focussing on groups and fields is done in [39].

Chapter 5

FORKING

Given two sets $A \subseteq B$ and a type $p \in S(A)$, we want to define a notion of a free extension of p to a type over B. When $A = M$ is a model, the work done in Chapter 2 provides us with a natural option: an extension of p is free if it extends to a global M-invariant type. However, it is not immediately clear how to generalize this to arbitrary base sets.

The idea of forking is to give such a definition (or rather a definition of the opposite: the free extensions are the non-forking ones). This definition makes sense in any theory, and satisfies different kinds of properties depending on the characteristics of the theory. For example, in *simple* theories, non-forking is an *independence relation*, which means in particular that it satisfies symmetry and transitivity (see for example [21]). In NIP theories, this is not true, but other properties hold. For instance, a type has only boundedly many non-forking extensions and over models non-forking coincides with invariance. Over an arbitrary set A, it coincides with bdd(A)-invariance, which we will define below.

Even though the general definition of forking is equivalent, for NIP theories, with an apparently simpler one, experience shows that it is useful to come back to it from time to time.

5.1. Bounded equivalence relations

In this section, we do not assume NIP. We consider relations on α-tuples, where α is a fixed finite or infinite ordinal.

An equivalence relation E between α-tuples of $\mathcal{U}$ is *A-invariant* if whenever $ab \equiv_A a'b'$, then $aEb \iff a'Eb'$. The relation xEy is type-definable over A if it is defined by a partial type $\pi(x; y)$ over A.

PROPOSITION 5.1. (*T any theory.*)

Let $A \subset \mathcal{U}$ and let E be an A-invariant equivalence relation on α-tuples of $\mathcal{U}$. Then the following are equivalent:

(i) *the set $\mathcal{U}^\alpha/E$ of E-equivalence classes is bounded (i.e., of size $< \bar{\kappa}$);*

(ii) $|\mathcal{U}^\alpha/E| \leq 2^{|A|+|T|}$;
(iii) *for any A-indiscernible sequence of α-tuples $(a_i : i < \omega)$ and $i, j < \omega$, we have $a_i E a_j$;*
(iv) *for any model $M \supseteq A$ and $a \equiv_M b$, we have aEb.*

PROOF. (ii) $\Rightarrow$ (i): Clear.

(iv) $\Rightarrow$ (ii): Clear by taking M of size $|A| + |T|$.

(iii) $\Rightarrow$ (iv): Let $a \equiv_M b$ and take p to be a coheir of $\text{tp}(a/M)$. In particular, p is M-invariant. Let $(a_i : i < \omega)$ be a Morley sequence of p over Mab. Then both $(a) + (a_i : i < \omega)$ and $(b) + (a_i : i < \omega)$ are indiscernible sequences. Therefore by (iii), we have aEa_0Eb.

(i) $\Rightarrow$ (iii): Assume there is some A-indiscernible sequence $(a_i : i < \omega)$ such that $\neg(a_i E a_j)$ for $i \neq j$. Then we may increase the sequence to one of size $\bar{\kappa}$. This contradicts (i). ⊣

An A-invariant equivalence relation E is *bounded* if it satisfies one of the equivalent conditions above.

DEFINITION 5.2. We say that two α-tuples a and b have the same Lascar strong type (or are Lascar-equivalent) over A if we have aEb for any A-invariant bounded equivalence relation E (on tuples of the right length). We write $\text{Lstp}(a/A) = \text{Lstp}(b/A)$.

LEMMA 5.3. *Lascar-equivalence over A is the finest bounded A-invariant equivalence relation (on tuples of a fixed length α).*

It is the transitive closure of the relation $\Pi_A(a; b)$ defined to hold if there is a model $M \supseteq A$ such that $a \equiv_M b$.

It is also the transitive closure of the relation $\Theta_A(a; b)$ defined to hold if (a, b) is the beginning of some infinite A-indiscernible sequence.

PROOF. It is clear from the definition that Lascar-equivalence over A is A-invariant and finer than any A-invariant bounded equivalence relation. It is bounded since for example it satisfies property (iii) of Proposition 5.1 (being the intersection of equivalence relations which satisfy it).

Let $\Pi_A^*(a; b)$ be the transitive closure of the relation $\Pi_A(x; y)$. It is an A-invariant equivalence relation and by Proposition 5.1 (iv), it is bounded. Furthermore, if $\Pi_A(a; b)$ holds, then again by Proposition 5.1 (iv) we have $\text{Lstp}(a/A) = \text{Lstp}(b/A)$. Hence $\Pi_A^*(a; b)$ implies $\text{Lstp}(a/A) = \text{Lstp}(b/A)$ and the two relations coincide.

The same argument goes through with Θ_A instead of Π_A using Proposition 5.1 (iii). ⊣

Note that if a and b are α-tuples and $\text{Lstp}(a/A) = \text{Lstp}(b/A)$ then we also have $\text{Lstp}(a_0/A) = \text{Lstp}(b_0/A)$ where a_0 is any subtuple of a and b_0 is the corresponding subtuple of b. However, one must beware that contrary to types, Lascar strong type are truly infinitary: Knowing that any two finite subtuples of a and b are Lascar-equivalent does not in general imply that a and b are Lascar-equivalent themselves. This will be apparent in the discussion below.

DEFINITION 5.4. We say that two α-tuples a and b are at Lascar-distance $\leq n$ over A if there exists a sequence $(a_i : i \leq n)$ of α-tuples such that $a_0 = a$, $a_n = b$ and we have $\Theta_A(a_i; a_{i+1})$ for every $i < n$. Two tuples are at Lascar-distance n if they are at Lascar-distance $\leq n$ but not $\leq n - 1$.

It follows from Lemma 5.3 that two tuples a and b are at finite Lascar-distance over A if and only if they are Lascar-equivalent over A.

Fix the parameter set A for this discussion and let $d_{\leq n}(x; y)$ be the relation saying that x and y are at Lascar-distance $\leq n$ over A. As $d_{\leq 1}(x; y)$ coincides with $\Theta(x; y)$, it is type-definable. It follows that each $d_{\leq n}(x; y)$ is type-definable and thus for a fixed n, we have $d_{\leq n}(a; b)$ if and only if we have $d_{\leq n}(a_0; b_0)$ for any $a_0 \subseteq a$ finite and $b_0 \subseteq b$ the corresponding subtuple in b. Hence we have shown the following lemma.

LEMMA 5.5. *Let a and b be two α-tuples. Then a and b are Lascar-equivalent over A if and only if there is some bound n such that we have $d_{\leq n}(a_0; b_0)$ for any $a_0 \subseteq a$ finite and $b_0 \subseteq b$ the corresponding subtuple in b.*

Example 5.6. For $n > 0$, let M_n be the structure with universe the standard unit circle equipped with a relation $R_n(x; y)$ which holds if and only if the arc between x and y has length at most $2\pi/n$. Then any two points in M_n are Lascar-equivalent and two diametrically opposite points have Lascar-distance $\lfloor n/2 \rfloor + 1$. Let M be the many-sorted structure $(M_n : n < \omega)$ and for each n, take two diametrically opposite points a_n, b_n in M_n. Then for each n, $(a_1, \dots, a_n)$ is Lascar-equivalent to $(b_1, \dots, b_n)$, but $(a_i : i < \omega)$ is not Lascar-equivalent to $(b_i : i < \omega)$.

We let $\mathrm{Autf}(\mathcal{U}/A)$ be the set of automorphisms of $\mathcal{U}$ generated by the groups $\mathrm{Aut}(\mathcal{U}/M)$ for M a model containing A.

LEMMA 5.7. *Two tuples a and b have the same Lascar strong type over A if and only if there is $\sigma \in \mathrm{Autf}(\mathcal{U}/A)$ such that $\sigma(a) = b$.*

PROOF. This follows at once from Lemma 5.3. ⊣

Exercise 5.8. The group $\mathrm{Autf}(\mathcal{U}/A)$ is exactly the subgroup of $\mathrm{Aut}(\mathcal{U})$ fixing all Lascar-strong types over A, that is the subgroup of automorphisms σ such that $\mathrm{Lstp}(a/A) = \mathrm{Lstp}(\sigma(a)/A)$ for any (finite or infinite) tuple a.

DEFINITION 5.9. We say that a global type p is Lstp_A-invariant if it is invariant under $\mathrm{Autf}(\mathcal{U}/A)$.

Hence a global type p is Lstp_A-invariant if and only if whenever a and b have same Lascar strong type over A and $\phi(x; y) \in L$, we have $p \vdash \phi(x; a) \leftrightarrow \phi(x; b)$. Equivalently and using Lemma 5.3, if and only if for any

A-indiscernible sequence $(a_i : i < \omega)$ and $d \models p$, the sequence $(a_i : i < \omega)$ is indiscernible over Ad.

DEFINITION 5.10. We say that two tuples a and b have the same compact strong type over A (or Kim-Pillay strong type), written $\text{KP-stp}(a/A) = \text{KP-stp}(b/A)$, if aEb holds for every bounded type-definable over A equivalence relation E (on tuples of the right length).

As in the case of Lascar strong types, we see that "equality of compact strong types over A" is the finest type-definable over A bounded equivalence relation.

We let $\text{Aut}(\mathcal{U}/\operatorname{bdd}(A))$ be the set of automorphisms σ of $\mathcal{U}$ such that $\text{KP-stp}(a/A) = \text{KP-stp}(\sigma(a)/A)$ for all a. We say that a global type p is $\operatorname{bdd}(A)$-invariant if it is invariant under $\text{Aut}(\mathcal{U}/\operatorname{bdd}(A))$.

Note that Lascar strong types refine compact strong types which refine types. If $A = M$ is a model, then by Lemma 5.3 two tuples having the same type over A have the same Lascar strong type over A, hence all those notions coincide.

PROPOSITION 5.11. *Let p be a type over A and let $R(x; y)$ be an A-invariant relation whose restriction to $p(\mathcal{U})^2$ is an equivalence relation with boundedly many classes, then:*

(i) *there is an A-invariant bounded equivalence relation E whose restriction to $p(\mathcal{U})^2$ coincides with R;*
(ii) *if furthermore R is type-definable, then there is a type-definable (over A) bounded equivalence relation E whose restriction to $p(\mathcal{U})^2$ refines R.*

PROOF. Point (i) is easy: simply take $E(x; y)$ to be $(R(x; y) \wedge p(x) \wedge p(y)) \vee (\neg p(x) \wedge \neg p(y))$.

Assume now that R is type-definable. Let $(a_i : i < \gamma)$ be a family of representatives of each class of $p(\mathcal{U})/R$. Let $E(x; y)$ be the relation stating that there are $(a'_i, a''_i : i < \gamma)$, all satisfying p such that $R(a'_i, a_i) \wedge R(a''_i, a_i)$ holds for each $i < \gamma$ and $\text{tp}(x(a'_i)_{i<\gamma}/A) = \text{tp}(y(a''_i)_{i<\gamma}/A)$. Then E is type-definable over $A(a_i)_{i<\gamma}$. Choosing another set of representatives yields the same relation E, therefore E is A-invariant and thus type-definable over A. It is clear that E is reflexive and symmetric. To see that it is transitive, assume that we have $E(x; y)$ witnessed by $(b_i, b'_i : i < \gamma)$ and $E(y; z)$ witnessed by $(c_i, c'_i : i < \gamma)$. Then there are $(d_i : i < \gamma)$ such that $\text{tp}(x(b_i)_{i<\gamma}(d_i)_{i<\gamma}/A) = \text{tp}(y(b'_i)_{i<\gamma}(c_i)_{i<\gamma}/A)$, thus $(d_i, c'_i : i < \omega)$ witnesses $E(x; z)$. Therefore E is an equivalence relation. If x and y have the same type over $A(a_i)_{i<\gamma}$, then they are E-equivalent (take $a'_i = a''_i = a_i$), therefore E is bounded. Finally, if x and y satisfy p and are E-equivalent, then this is witnessed by some $(a'_i, a''_i : i < \gamma)$. For some $i < \gamma$, we have $R(x; a'_i)$, therefore also $R(y; a''_i)$ holds and then $R(x; y)$ holds. Thus E has the required properties. ⊣

5.2. Forking

In this section, we again make no assumption on the theory T.

DEFINITION 5.12. Let $A \subset \mathcal{U}$.

(i) A formula $\phi(x;b) \in L(\mathcal{U})$ *divides* over A if there is an A-indiscernible sequence $(b_i : i < \omega)$ where $b_0 = b$ such that the partial type $\{\phi(x;b_i) : i < \omega\}$ is inconsistent.

(ii) A partial type $\pi(x)$ *divides* over A if it implies a formula which divides over A;

(iii) A partial type $\pi(x)$ *forks* over A if it implies a finite disjunction $\bigvee_{i<n} \phi_i(x;b^i)$ of formulas, such that $\phi_i(x;b^i)$ divides over A for each $i < n$.

As the next lemma shows, dividing is a property of a definable set and does not depend on the formula used to define it.

LEMMA 5.13. *If $\phi(x;b)$ implies a formula $\psi(x;c)$ which divides over A, then $\phi(x;b)$ divides over A.*

PROOF. Let $(c_i : i < \omega)$ be A-indiscernible with $c_0 = c$. Then one can extend it to a sequence $(b_i\hat{~}c_i : i < \omega)$ which is A-indiscernible and where $b_0 = b$. In particular, we have $\models \phi(x;b_i) \rightarrow \psi(x;c_i)$ for all i. If the partial type $\{\psi(x;c_i) : i < \omega\}$ is inconsistent, then so is $\{\phi(x;b_i) : i < \omega\}$. ⊣

The advantage of forking over dividing lies in point 4 of the following proposition.

PROPOSITION 5.14. 1. *If $\pi(x)$ is a consistent partial type over A, then it does not divide over A.*

2. *If the partial type $\pi(x)$ divides (resp. forks) over A, then some finite $\pi_0(x) \subseteq \pi(x)$ divides (resp. forks) over A.*
3. *Let $A \subset M$, M is $|A|^+$-saturated and let $p(x)$ be a complete type over M. Then p does not fork over A if and only if it does not divide over A.*
4. *Let $A \subseteq B$ and $\pi(x)$ be a partial type over B, then $\pi(x)$ does not fork over A if and only if it has an extension to a global complete type which does not divide (equiv. fork) over A.*
5. *If $A \subseteq B \subseteq C$ and $a \in \mathcal{U}$, then if* $\mathrm{tp}(a/C)$ *does not divide (resp. fork) over A, it does not divide (resp. fork) over B and* $\mathrm{tp}(a/B)$ *does not divide (resp. fork) over A.*
6. *If $p(x)$ is a global* Lstp_A*-invariant type, then it does not divide (equiv. fork) over A.*

PROOF. 1. If $\pi(x)$ implies a formula $\phi(x;b_0)$ and $(b_i : i < \omega)$ is an A-indiscernible sequence, then $\pi(x)$ implies all formulas $\phi(x;b_i)$, $i < \omega$. As $\pi(x)$ is consistent, so is $\{\phi(x;b_i) : i < \omega\}$ and it follows that $\phi(x;b_0)$ does not divide over A.

2. Clear by compactness.

3. Assume that $p(x)$ implies a finite disjunction of formulas $\bigvee_{i<n}\phi_i(x;b)$, each of which divides over A. There is some finite $C_0 \subseteq M$ such that $p'(x) = p(x)|_{C_0}$ already implies this disjunction. Let $C = C_0 \cup A$. We may find a tuple $b' \in M$, $b' \equiv_C b$. Then $p'(x)$ implies $\bigvee_{i<n}\phi_i(x;b')$, hence one of the formulas $\phi_i(x;b')$ is in $p(x)$, and $p(x)$ divides over A.
4. Let $\pi(x)$ be a partial type over A. Let $\Sigma(x)$ be the set of formulas $\neg\phi(x;b)$ where $\phi(x;b) \in L(\mathcal{U})$ divides over A. If $\pi(x) \cup \Sigma(x)$ is consistent, then it extends to a complete type over $\mathcal{U}$ which does not divide over A. Otherwise, by compactness, there is some finite part $\Sigma_0(x)$ of $\Sigma(x)$ such that $\pi(x) \cup \Sigma_0(x)$ is inconsistent. Hence $\pi(x)$ forks over A.
5. Clear from the definition.
6. Assume that $p(x) \vdash \phi(x;b)$ for some $\phi(x;b) \in L(\mathcal{U})$, and let $(b_i : i < \omega)$ be an A-indiscernible sequence with $b_0 = b$. Then as $p(x)$ is Lstp_A-invariant, $p(x) \vdash \phi(x;b_i)$ for each $i < \omega$. As p is consistent, so is the conjunction $\bigwedge_{i<\omega}\phi(x;b_i)$. ⊣

Example 5.15. We give an example of a type $p \in S(A)$ which forks over A.

Let $\mathbb{U}$ be the usual unit circle in the plane and define a ternary relation $R(x, y, z)$ on $\mathbb{U}$ which holds if and only if either x and z are diametrically opposite, or y lies in the (closed) small arc between x and z. Consider the structure $M = (\mathbb{U}; R)$. Let $p \in S_1(\emptyset)$ be the unique 1-type over $\emptyset$. We claim that p forks over $\emptyset$.

To see this, let $b, c, d \in \mathbb{U}$ divide the circle into three small arcs. Then $p \vdash R(b, x, c) \vee R(c, x, d) \vee R(d, x, b)$. It is enough to show that say $R(b, x, c)$ divides over $\emptyset$. We can find an indiscernible sequence $(b_i, c_i : i < \omega)$ such that $(b_0, c_0) = (b, c)$ and the points $b_0, c_0, b_1, c_1, \dots$ lie in that order on the circle. Then the formula $R(b_0, x, c_0) \wedge R(b_1, x, c_1)$ is inconsistent as required.

Note that this theory is NIP, since it is interpretable in RCF.

DEFINITION 5.16. An *extension base* is a set $A \subset \mathcal{U}$ such that no $p \in S(A)$ forks over A.

LEMMA 5.17. *Let $A, b \subset \mathcal{U}$ and π a partial type over Ab. Then the following are equivalent*:

(i) *π does not divide over A;*

(ii) *for every A-indiscernible sequence $I = (b_i : i < \omega)$ with $b_0 = b$, there is $a \models \pi$ such that I is indiscernible over Aa.*

PROOF. It is clear that (ii) implies (i).

Conversely, assume (i) and let $I = (b_i : i < \omega)$ with $b_0 = b$ be an A-indiscernible sequence. Write $\pi(x) = \pi(x; b_0)$ where π has hidden parameters from A. By the non-dividing assumption, the partial type $\bigcup\{\pi(x;b_i) : i < \omega\}$ is consistent. Let a' realize it. By Ramsey, we may find an Aa'-indiscernible sequence $(b_i' : i < \omega)$ realizing the EM-type of $(b_i : i < \omega)$ over Aa'. Then $\mathrm{tp}(b_0'/Aa') = \mathrm{tp}(b_0/Aa')$ so without loss $b_0' = b_0$. Let $f \in \mathrm{Aut}(\mathcal{U}/Ab)$ send

$(b'_i : i < \omega)$ to $(b_i : i < \omega)$ and set $a = f(a')$. Then the sequence I is indiscernible over Aa. ⊣

LEMMA 5.18. *Let $A \subseteq B \subset \mathcal{U}$ and $a, b \in \mathcal{U}$. Assume that* $\text{tp}(a/B)$ *does not fork over A and* $\text{tp}(b/Ba)$ *does not fork over Aa, then* $\text{tp}(a, b/B)$ *does not fork over A.*

PROOF. Note that during the proof, we may freely replace (a, b) by any other pair which has the same type over B.

Let M be some $|B|^+$-saturated model containing B. As the type $\text{tp}(a/B)$ does not fork over A, it has an extension p' over M which does not fork over A. Let a' realize p'. There is an automorphism σ fixing B and sending a to a'. Replacing (a, b) by $(\sigma(a), \sigma(b))$ we may assume that $\text{tp}(a/M)$ does not fork over A. Similarly, as the type $\text{tp}(b/Ba)$ does not fork over Aa, it has an extension q' to a type over Ma which does not fork over Aa. Replacing b by a realization of q', we may assume that $\text{tp}(b/Ma)$ does not fork over Aa.

By Proposition 5.14 (3), it is enough to prove that $\text{tp}(a, b/M)$ does not divide over A. So assume that it does divide and take $\phi(x, y; c) \in \text{tp}(a, b/M)$ and an A-indiscernible sequence $(c_i : i < \omega)$, $c_0 = c$ such that $\{\phi(x, y; c_i) : i < \omega\}$ is inconsistent. By Lemma 5.17, there is an automorphism f fixing Ac pointwise such that $(c_i : i < \omega)$ is indiscernible over $f(a)$. Then $\text{tp}(f(b)/Af(a)c)$ does not divide (indeed does not fork) over $Af(a)$. Therefore there is an automorphism g fixing $Af(a)c$ such that $(f(a)c_i : i < \omega)$ is indiscernible over $g(f(b))$. We then have $\models \phi(f(a), g(f(b)); c_i)$ for all $i < \omega$. Contradiction. ⊣

COROLLARY 5.19. *Assume that for every $A \subset \mathcal{U}$ and $p \in S_1(A)$, p does not fork over A, then every A is an extension base.*

PROOF. This follows immediately from the previous lemma by induction on the arity of p. ⊣

COROLLARY 5.20. *Let $p(x)$ and $q(y)$ be two global invariant types, both non-forking over A. Then $p(x) \otimes q(y)$ is non-forking over A.*

PROOF. It is enough to show that $p(x) \otimes q(y)|_B$ is non-forking over A for any small B containing A. Take such a B; let $b \models q|_B$ and $a \models p|_{Bb}$. Then $\text{tp}(b/B)$ does not fork over A, nor does $\text{tp}(a/Bb)$. Hence by the previous lemma, $\text{tp}(a, b/B) = p(x) \otimes q(y)|_B$ does not fork over A. ⊣

PROPOSITION 5.21. *(T is NIP.) Let $A \subset \mathcal{U}$, and let $p \in S(\mathcal{U})$ be a global type. Then p does not fork over A if and only if it is* Lstp_A*-invariant.*

PROOF. Right to left has already been observed and holds in any theory.

We show left to right. Let $p \in S(\mathcal{U})$ and assume that p is not Lstp_A-invariant. This implies that there is some A-indiscernible sequence $(c_i : i < \omega)$ and some formula $\phi(x; y)$ such that $p \vdash \phi(x; c_0) \wedge \neg\phi(x; c_1)$. Consider the A-indiscernible sequence of pairs $(c_{2i}c_{2i+1} : i < \omega)$. The partial type

$\{\phi(x; c_{2i} \wedge \neg\phi(x; c_{2i+1})\}$ is inconsistent by NIP. This implies that the formula $\phi(x; c_0) \wedge \neg\phi(x; c_1)$ divides over A. ⊣

COROLLARY 5.22. *(T is NIP.) Let $M \prec \mathcal{U}$, and p be a global type. Then p does not fork over M if and only if p is M-invariant.*

Let $p \in S(\mathcal{U})$ be a global type. If p is M-invariant for some small M, then it does not fork over M. Conversely, if p does not fork over some small $A \subset \mathcal{U}$, then it is M-invariant for any model M containing A. So for a global type p the properties "p is invariant over some small $M \prec \mathcal{U}$" and "p does not fork over some small $A \subset \mathcal{U}$" are equivalent.

COROLLARY 5.23. *(T is NIP.) Let $p \in S(A)$ be a type in a finite number of variables. Then p has at most $2^{|A|+|T|}$ non-forking global extensions.*

PROOF. Take M a model containing A such that $|M| = |A| + |T|$. If q is a non-forking extension of p, then q does not fork over M, therefore it is M-invariant. We have seen in the proof of Proposition 2.43 that there are at most $2^{|M|}$ global M-invariant types. ⊣

Readers familiar with simple theories know that forking satisfies additional properties in that context, in particular symmetry (if $\text{tp}(a/Ab)$ forks over A, then so does $\text{tp}(b/Aa)$) and transitivity (if $A \subseteq B \subseteq C$, $\text{tp}(a/B)$ does not fork over A and $\text{tp}(a/C)$ does not fork over B, then $\text{tp}(a/C)$ does not fork over A). It is known that each one of these properties is actually equivalent to simplicity of the theory, hence will hold in no unstable NIP theory. We give a concrete example.

Example 5.24 (Failure of symmetry and transitivity). Let M be a model of DLO and let p be any non-algebraic 1-type over M. Let $a < b < c$ all realize p. Then $\text{tp}(b/Mac)$ forks over M, indeed an indiscernible sequence $(a_i, c_i : i < \omega)$ of realizations of p with $a_i < c_i < a_{i+1}$ witnesses dividing. However $\text{tp}(ac/Mb)$ does not fork over M since this type extends to a global M-invariant type.

We obtain a failure of transitivity by observing that $\text{tp}(b/Mac)$ does not fork over Ma, $\text{tp}(b/Ma)$ does not fork over M, but $\text{tp}(b/Mac)$ does fork over M.

5.3. bdd(A)-invariance

Assumption: In this section, we assume that T is NIP.

PROPOSITION 5.25. *Let $p \in S(\mathcal{U})$ be non-forking over A. Then the equivalence relation $\text{Lstp}(x/A) = \text{Lstp}(y/A)$ restricted to realizations of $p|_A$ is type-definable.*

Indeed, if $c, d \models p|_A$, then: $\text{Lstp}(c/A) = \text{Lstp}(d/A)$ if and only if there is some sequence I such that both $(c) + I$ and $(d) + I$ realize $p^{(\omega)}|_A$.

PROOF. It is enough to prove the second statement.

($\Leftarrow$) Clear; the condition implies that c and d are at Lascar-distance at most 2.

($\Rightarrow$) Let c, d realize $p|_A$, and assume that $\operatorname{Lstp}(c/A) = \operatorname{Lstp}(d/A)$. Let M be a model containing A. The hypothesis and conclusion depend only on $\operatorname{tp}(c, d/A)$, hence we may conjugate (c, d) by an automorphism over A and assume that $\operatorname{tp}(c/M) = p|_M$. We let $I \models p^{(\omega)}|_{Mcd}$. Then $c + I$ is a realization of $p^{(\omega)}|_A$ (indeed of $p^{(\omega)}|_M$). By Corollary 5.20, $p^{(\omega)}$ does not fork over A. By Proposition 5.21, it is Lascar-invariant over A, therefore $\operatorname{tp}(cI/A) = \operatorname{tp}(dI/A)$ and both $(c) + I$ and $(d) + I$ realize $p^{(\omega)}|_A$. ⊣

COROLLARY 5.26. *If $p \in S(A)$ does not fork over A and $a, b \models p$ satisfy* $\text{KP-stp}(a/A) = \text{KP-stp}(b/A)$, *then* $\operatorname{Lstp}(a/A) = \operatorname{Lstp}(b/A)$.

In particular, if no type forks over its base, then compact strong types coincide with strong types. (We say that T is G-compact.)

PROOF. Let $p \in S(A)$ be non-forking over A. By the previous proposition, the relation $R(c; d)$ defined as $\operatorname{Lstp}(c/A) = \operatorname{Lstp}(d/A)$ is type-definable on realizations of p. By Proposition 5.11, there is a type-definable bounded equivalence relation E which refines R on p (and must therefore be equal to R by definition of Lascar strong types). If $c, d \models p$ have the same KP-strong type, then they are E-equivalent and thus R-equivalent. ⊣

We now prove an analog of Proposition 2.36 for types non-forking over A. First a lemma.

LEMMA 5.27. *Let p be a global type, non-forking over A. Let $b_1, b_2 \in \mathcal{U}$ have the same Lascar strong type over A and let $a \models p \upharpoonright Ab_1b_2$. Then* $\operatorname{Lstp}(ab_1/A) = \operatorname{Lstp}(ab_2/A)$.

PROOF. We might as well take a to realize p over $\mathcal{U}$. First assume that there is a model $M \supseteq A$ such that $\operatorname{tp}(b_1/M) = \operatorname{tp}(b_2/M)$. Then p is M-invariant and we have $\operatorname{tp}(ab_1/M) = \operatorname{tp}(ab_2/M)$, which implies $\operatorname{Lstp}(ab_1/A) = \operatorname{Lstp}(ab_2/M)$. In general there is a finite sequence of tuples $(c_i : i \leq n)$ and a sequence $(M_i : i < n)$ of models containing A such that $c_0 = b_1$, $c_n = b_2$ and $\operatorname{tp}(c_i/M_i) = \operatorname{tp}(c_{i+1}/M_i)$ for each $i < n$. Then $\operatorname{Lstp}(ac_i/A) = \operatorname{Lstp}(ac_{i+1}/A)$ for each i and the result follows. ⊣

PROPOSITION 5.28. *Let $p, q \in S(\mathcal{U})$ be two global types, both non-forking over A and let $\bar{a} \models p^{(\omega)}$ and $\bar{b} \models q^{(\omega)}$. Assume that* $\operatorname{Lstp}(\bar{a}/A) = \operatorname{Lstp}(\bar{b}/A)$, *then $p = q$.*

PROOF. The proof is essentially the same as that of Proposition 2.36, replacing types over A by Lascar strong types over A.

Assume that there is some $\phi(x; d) \in L(\mathcal{U})$ such that $p \vdash \phi(x; d)$ and $q \vdash \neg\phi(x; d)$. Pick any model $M \supseteq A \cup \{d\}$. We build inductively two sequences $(a_n : n < \omega)$ and $(b_n : n < \omega)$ such that for all n:

- $a_n \models p \upharpoonright Ma_{<n}b_{<n}$;

- $b_n \models q \upharpoonright M a_{\leq n} b_{<n}$.

It is enough now to show that the sequence $(a_0, b_0, a_1, b_1, \dots)$ is indiscernible, because the formula $\phi(x; d)$ alternates infinitely often on that sequence, contradicting NIP.

We show by induction on n that $I_n := (a_0, b_0, \dots, a_n)$ has the same Lascar strong type over A as a realization of $q^{(2n+1)}$. For $n = 0$, it follows from the hypothesis of the proposition. Assume we know it for n. Then by Lemma 5.27 and induction hypothesis, $I_n + (b_n)$ has the same Lascar strong type over A as $q^{(2n+2)}$. By hypothesis, this is the same as the Lascar strong type of $p^{(2n+2)}$. We then apply the same argument to add a_{n+1} and this finishes the induction. $\dashv$

COROLLARY 5.29. *Let $p \in S(\mathcal{U})$ be non-forking over A, then p is* bdd(A)-*invariant.*

PROOF. Let $\sigma \in \mathrm{Aut}(\mathcal{U}/\,\mathrm{bdd}(A))$, $\bar{a} \models p^{(\omega)}|_M$. Then

$$\text{KP-stp}(\bar{a}/A) = \text{KP-stp}(\sigma(\bar{a})/A).$$

As $\bar{a}$ and $\sigma(\bar{a})$ are both realizations of $p^{(\omega)}|_A$, which is non-forking over A, Corollary 5.26 implies that $\mathrm{Lstp}(\bar{a}/A) = \mathrm{Lstp}(\sigma(\bar{a})/A)$. Therefore by Proposition 5.28, $p = \sigma(p)$. $\dashv$

5.4. NTP$_2$ and the broom lemma

DEFINITION 5.30. (T any theory.) We say that a formula $\phi(x; y)$ has TP$_2$ if there is an array $(b_i^t : i < \omega, t < \omega)$ of tuples of size $|y|$ and $k < \omega$ such that:

- for any $\eta : \omega \to \omega$, the conjunction $\bigwedge_{t<\omega} \phi(x; b^t_{\eta(t)})$ is consistent;
- for any $t < \omega$, $\{\phi(x; b_i^t) : i < \omega\}$ is k-inconsistent.

We say that $\phi(x; y)$ is NTP$_2$ if it does not have TP$_2$. We say that the theory T is NTP$_2$ if all formulas are NTP$_2$.

PROPOSITION 5.31. *If T is NIP, then it is NTP$_2$.*

PROOF. Assume that some formula $\phi(x; y)$ has TP$_2$, as witnessed by some $k < \omega$ and array $(a_i^t : i < \omega, t < \omega)$. By compactness, we may increase the array to $(a_i^t : i < \omega, t < |T|^+)$. Next, using Lemma 4.2, we may assume that the rows $(a_i^t)_{i<\omega}$ are mutually indiscernible. Take b realizing the partial type $\{\phi(x; a_0^t) : t < |T|^+\}$. Then by k-inconsistency of the rows, no sequence $(a_i^t)_{i<\omega}$ remains indiscernible over b. This contradicts Lemma 4.8. $\dashv$

The class of NTP$_2$ theories contains both NIP theories and simple theories. Most of the content of this section and the next one goes through (with some modifications) for NTP$_2$ theories. However, we do not try to be optimal here, hence:

Assumption: Until the end of the chapter, we assume that T is NIP.

Notation 5.32. $a \mathop{\smile\!\!\!\!\!\!\:|}_A b$ means that $\mathrm{tp}(a/Ab)$ does not fork over A.

Beware that this is not in general a symmetric relation.

DEFINITION 5.33. Let $I = (a_i : i < \omega)$ be a sequence of tuples. We say that it is a *Morley sequence* over A if there is some global type p, non-forking over A such that $I \models p^{(\omega)}|_A$.

LEMMA 5.34. *Let $\bar{b} = (b_i : i < \omega)$ be an A-indiscernible sequence such that $p = \mathrm{tp}(\bar{b}/A)$ does not fork over A. Then there is some global non-forking extension $q(x_0, x_1, \dots)$ of p which is the type of some $\mathcal{U}$-indiscernible sequence.*

PROOF. Let p' be any non-forking extension of $\mathrm{tp}(\bar{b}/A)$. Let $(d_i : i < \omega) \models p'$. We can find some $\mathcal{U}$-indiscernible sequence $(e_i : i < \omega)$ realizing the EM-type of $(d_i : i < \omega)$ over $\mathcal{U}$. Let $q = \mathrm{tp}(e_i : i < \omega/\mathcal{U})$. Then q does not fork over A and its restriction to A is p. ⊣

LEMMA 5.35. *Let A be an extension base, and $\phi(x; b) \in L(\mathcal{U})$ divide over A. Then there is a Morley sequence $(b_i : i < \omega)$ over A witnessing dividing (i.e., $b_0 = b$ and the conjunction $\bigwedge_{i<\omega} \phi(x; b_i)$ is inconsistent).*

PROOF. Let $(b'_i : i < \omega)$ be some A-indiscernible sequence with $b'_0 = b$ and such that $\{\phi(x; b'_i) : i < \omega\}$ is k-inconsistent. Let $p = \mathrm{tp}((b'_i)_{i<\omega}/A)$. By assumption on A, p does not fork over A, so by the previous lemma, there is some $q(x_0, \dots) \in S(\mathcal{U})$ a non-forking extension of p and some $r(x) \in S(\mathcal{U})$ such that the type q restricted to any one of its variable is equal to r. Let $(\bar{b}^t : t < \omega)$ be a Morley sequence of q over A, where $\bar{b}^t = (b^t_i : i < \omega)$.

For each $t < \omega$, the type $\{\phi(x; b^t_i) : i < \omega)\}$ is k-inconsistent. By NTP$_2$, there is some $\eta : \omega \to \omega$ such that $\bigwedge_{t<\omega} b^t_{\eta(t)}$ is inconsistent. As the sequence $(b^t_{\eta(t)} : t < \omega)$ is a Morley sequence of r over A, we have what we want. ⊣

LEMMA 5.36 (Broom lemma). *Let $A \subset \mathcal{U}$ be an extension base and $\pi(x)$ be a partial type over $\mathcal{U}$ which is Lstp_A-invariant. Assume that for some formulas $\psi(x; b)$ and $\phi_i(x; c)$, $i < n$ in $L(\mathcal{U})$ we have*

$$\pi(x) \vdash \psi(x; b) \vee \bigvee_{i<n} \phi_i(x; c),$$

where $b \mathop{\smile\!\!\!\!\!\!\:|}_A c$ and for each $i < n$, $\phi_i(x; c)$ divides over A. Then

$$\pi(x) \vdash \psi(x; b).$$

PROOF. We prove the result by induction on n. The case $n = 0$ is trivial.

Assume we know it for n and let $\pi(x) \vdash \psi(x; b) \vee \bigvee_{i<n+1} \phi_i(x; c)$. By the previous lemma, there is a Morley sequence $(c_j : j < \omega)$ over A with $c_0 = c$ and such that $\{\phi_n(x; c_j) : j < \omega\}$ is k-inconsistent. Conjugating by an automorphism, we may assume that $b \mathop{\smile\!\!\!\!\!\!\:|}_A (c_j)_{j<\omega}$, thus the sequence $(c_j)_{j<\omega}$ is indiscernible over Ab (since $\mathrm{tp}(b/A(c_j))$ is Lstp_A-invariant).

By Lstp_A-invariance of $\pi(x)$, we have $\pi(x) \vdash \psi(x;b) \vee \bigvee_{i<n+1} \phi_i(x;c_j)$ for all $j < \omega$. In particular, we have

$$\pi(x) \vdash \psi(x;b) \vee \bigwedge_{j<k} \bigvee_{i<n+1} \phi_i(x;c_j).$$

By assumption on k, this implies

$$\pi(x) \vdash \psi(x;b) \vee \bigvee_{\substack{j<k \\ i<n}} \phi_i(x;c_j).$$

For any $j < k$, we have $b \mathop{\smile\!\!\!\!\!\!\,|}_A c_{\geq j}$, thus $b \mathop{\smile\!\!\!\!\!\!\,|}_{Ac_{>j}} c_j$. Also we have $c_{>j} \mathop{\smile\!\!\!\!\!\!\,|}_A c_j$. Therefore by left transitivity (Lemma 5.18) $bc_{>j} \mathop{\smile\!\!\!\!\!\!\,|}_A c_j$.

By the induction hypothesis, and as

$$\pi(x) \vdash \left[\psi(x;b) \vee \bigvee_{\substack{0<j<k \\ i<n}} \phi_i(x;c_j)\right] \vee \bigvee_{i<n} \phi_i(x;c_0)$$

we have

$$\pi(x) \vdash \psi(x;b) \vee \bigvee_{\substack{0<j<k \\ i<n}} \phi_i(x;c_j).$$

Iterating this last step $k-1$ times, we obtain $\pi(x) \vdash \psi(x;b)$. ⊣

The following corollary has the same hypothesis as the lemma, except that we drop the condition $b \mathop{\smile\!\!\!\!\!\!\,|}_A c$.

COROLLARY 5.37. *Let $A \subset \mathcal{U}$ be an extension base and $\pi(x)$ be a partial type over $\mathcal{U}$ which is* Lstp_A*-invariant. Assume that for some formulas $\psi(x;b)$ and $\phi_i(x;c)$, $i<n$ in $L(\mathcal{U})$ we have*

$$\pi(x) \vdash \psi(x;b) \vee \bigvee_{i<n} \phi_i(x;c),$$

and for each $i<n$, $\phi_i(x;c)$ divides over A.

Then we have $\pi(x) \vdash \bigvee_{j<m} \psi(x;b_j)$, where the tuples b_j are such that $\mathrm{Lstp}(b_j/A) = \mathrm{Lstp}(b/A)$.

PROOF. Consider the partial type

$$\pi'(x) = \pi(x) \cup \{\neg\psi(x;b') : \mathrm{Lstp}(b'/A) = \mathrm{Lstp}(b/A)\}.$$

Then $\pi'(x)$ is Lstp_A-invariant and by hypothesis on $\pi(x)$, we have $\pi'(x) \vdash x \neq x \vee \bigvee \phi_i(x;c)$. The previous lemma applies to give $\pi'(x) \vdash x \neq x$. Therefore $\pi'(x)$ is inconsistent and the conclusion follows by compactness. ⊣

DEFINITION 5.38. Let p be a type over a model M and $N \succ M$. An *heir* of p over N is a type $q \in S(N)$ which extends p and such that whenever $q \vdash \phi(x;b)$, with $\phi(x;y) \in L(M)$ and $b \in N$, then there is $b' \in M$ such that $p \vdash \phi(x;b')$.

Note that if $q \in S(N)$ is an heir of its restriction p to M and $a \models q$, then $\operatorname{tp}(N/Ma)$ is finitely satisfiable in M and in particular we have $N \mathop{\smile\!\!\!\!\!\!\mid}_M a$. It is a well known fact (that we will not use) that a type $p \in S(M)$ is definable if and only if it admits a unique global heir.

PROPOSITION 5.39. *Let $M \models T$ and $p \in S(M)$. Then p has a global heir which is non-forking over M.*

PROOF. Assume not. Then the partial type $\pi(x) = p(x) \cup \pi_1(x) \cup \pi_2(x)$ is inconsistent where

- $\pi_1(x) = \{\psi(x;b) \in L(\mathcal{U}) : p \vdash \psi(x;b') \text{ for all } b' \in M\}$ and
- $\pi_2(x) = \{\phi(x;c) \in L(\mathcal{U}) : \neg\phi(x;c) \text{ divides over } M\}$.

By compactness, we find a formula $\psi(x;b)$ from $\pi_1(x)$ and finitely many formulas $\phi_i(x;c)$, $i < n$ in $\pi_2(x)$ such that $p(x) \vdash \neg\psi(x;b) \vee \bigvee_{i<n} \neg\phi_i(x;c)$. By Corollary 5.37, we have $p(x) \vdash \bigvee_{j<m} \neg\psi(x;b_j)$, where $b_j \equiv_M b$. As p is a type over M, by compactness, there is some formula $\theta(y_0, \dots, y_{m-1}) \in \operatorname{tp}(b_0, \dots, b_{m-1}/M)$ such that

$$p(x) \vdash \forall y_0, \dots, y_{m-1}\left(\theta(\bar{y}) \to \bigvee_{j<m} \neg\psi(x;y_j)\right).$$

Taking such $y_0, \dots, y_{m-1}$ in M, we obtain a contradiction to the fact that $\psi(x;b)$ is in $\pi_1(x)$. ⊣

5.5. Strict non-forking

We continue to assume that T is NIP.

DEFINITION 5.40. Let $A \subseteq B$ and $a \in \mathcal{U}$. We say that $p = \operatorname{tp}(a/B)$ is *strictly non-forking* over A if there is a global extension p' of p such that p' is non-forking over A and such that for all $C \supseteq B$, if $a_0 \models p'|_C$, then $C \mathop{\smile\!\!\!\!\!\!\mid}_A a_0$.

We will write $a \mathop{\smile\!\!\!\!\!\!\mid}_A^{st} b$ to mean that $\operatorname{tp}(a/Ab)$ is strictly non-forking over A.

Remark 5.41. In particular, if $\operatorname{tp}(a/B)$ is strictly non-forking over A, then we have $a \mathop{\smile\!\!\!\!\!\!\mid}_A B$ and $B \mathop{\smile\!\!\!\!\!\!\mid}_A a$. It is an open question whether the converse holds (assuming that A is an extension base).

Example 5.42. Let M be a model of DLO. As in Example 5.24, take any non-algebraic type $p \in S_1(M)$ and $a < b < c$ realizing p. Then $\operatorname{tp}(a, c/Mb)$ does not fork over M. However it is not strictly non-forking because $\operatorname{tp}(b/Mac)$ forks over M. There are two strictly non-forking extensions of $\operatorname{tp}(a, c/M)$ to a type over Mb obtained by placing a and c on the same side of b.

Example 5.43. Let $p \in S(M)$, then p is strictly non-forking over M: By Proposition 5.39 and the remark before it we can take p' to be any global non-forking heir of p.

We now generalize this to types over arbitrary extension bases.

PROPOSITION 5.44. *Let A be an extension base and $p \in S(A)$. Then p is strictly non-forking over A.*

PROOF. Let $a \models p$. The proof is essentially the same as that of Proposition 5.39. Namely, we consider the partial type $p(x) \cup \{\psi(x;b) : b \in \mathcal{U}, \neg\psi(a;y) \text{ forks over } A\} \cup \{\phi(x;c) : c \in \mathcal{U}, \neg\phi(x;c) \text{ divides over } A\}$. If this type is inconsistent, then we can find some $\psi(x;b)$ from the first set and finitely many $\phi_i(x;c)$, $i < n$ from the second set such that

$$p(x) \vdash \neg\psi(x;b) \vee \bigvee_{i<n} \neg\phi_i(x;c).$$

By Corollary 5.37, we have $p(x) \vdash \bigvee_{j<m} \neg\psi(x;b_j)$ where $\mathrm{Lstp}(b_j/A) = \mathrm{Lstp}(b/A)$. Let $\bar{b} = b_0{}^\frown \dots {}^\frown b_{m-1}$ and take $q \in S(\mathcal{U})$ a non-forking extension of $\mathrm{tp}(\bar{b}/A)$ (as A is an extension base). If $\bar{b}' \models q|_{Aa}$, then by hypothesis on $\psi(x;y)$, $\psi(a;b_i)$ holds for all i. A contradiction. ⊣

DEFINITION 5.45. The sequence $(a_i : i < \alpha)$ is a *strict non-forking sequence* over A if for any $i < \alpha$ we have $a_i \mathop{\smile\!\!\!\!\!\mid}\nolimits^{st}_A a_{<i}$.

LEMMA 5.46. *Let A be an extension base. Let $(a_i : i < \alpha)$ be a strict non-forking sequence over A and $(I_i : i < \alpha)$ a sequence of A-indiscernible sequences such that I_i begins with a_i. Then we can find sequences $(J_i : i < \alpha)$ mutually indiscernible over A such that $J_i \equiv_{Aa_i} I_i$ for each $i < \alpha$.*

PROOF. Without loss, the sequences I_i are indexed by ω. We may assume that $\alpha = n < \omega$ and we show the result by induction on n. Assume it is true for n and let $(a_i : i < n+1)$ and $(I_i : i < n+1)$ be given. By the induction hypothesis, we may assume that the sequences $(I_i : i < n)$ are mutually indiscernible over A. We have $a_n \mathop{\smile\!\!\!\!\!\mid}\nolimits^{st}_A a_{<n}$. Conjugating by an automorphism fixing A and the family $(a_i : i < n+1)$, we may assume that $a_n \mathop{\smile\!\!\!\!\!\mid}\nolimits^{st}_A a_{<n} I_{<n}$. This implies:

(1) the sequences $(I_i : i < n)$ are mutually indiscernible over Aa_n;
(2) $a_{<n} I_{<n} \mathop{\smile\!\!\!\!\!\mid}\nolimits_A a_n$.

By Lemma 5.17, $\mathrm{tp}(a_{<n} I_{<n}/Aa_n)$ has an extension q over AI_n such that I_n is indiscernible over a realization of q. Therefore moving I_n by an automorphism over Aa_n, we may assume that I_n is indiscernible over $Aa_{<n} I_{<n}$.

For $k < n+1$, write $I_k = (a_i^k : i < \omega)$. Let $\eta : n+1 \to \omega$ be any function. As I_n is indiscernible over $AI_{<n}$, we have

$$a^0_{\eta(0)} \cdots a^n_{\eta(n)} \equiv_A a^0_{\eta(0)} \cdots a^{n-1}_{\eta(n-1)} a_n.$$

Then by (1) above,

$$a^0_{\eta(0)} \cdots a^n_{\eta(n)} \equiv_A a_0 \cdots a_n.$$

For k from 0 to $n-1$, construct a sequence I'_k which is indiscernible over $AI_n I_{>k} I'_{<k}$ and realizes the EM-type of I_k over that set. Then the sequences $(I'_k : k < n)$ are mutually indiscernible over AI_n. Write $I'_k = (a'^k_i : i < \omega)$. By the previous observation we have $a'^0_0 \cdots a'^{n-1}_0 \equiv_{Aa_n} a_0 \cdots a_{n-1}$. Therefore we may assume that $a'^k_0 = a_k$ for all $k < n$.

Then the sequences $I'_0, \ldots, I'_{n-1}, I_n$ have the required properties. ⊣

PROPOSITION 5.47. *Let A be an extension base, and let $(a_i : i < \kappa)$ be a strict non-forking sequence over A. Let b such that* dp-rk$(b/A) < \kappa$. *Then there is $i < \kappa$ such that* tp(b/Aa_i) *does not divide over A.*

PROOF. Assume that tp(b/Aa_i) divides over A for all $i < \kappa$. Then, for each $i < \kappa$ we can find some $\phi_i(x; a_i) \in \text{tp}(b/Aa_i)$ and an A-indiscernible sequence $I_i = (a^i_j : j < \omega)$ such that $a^i_0 = a_i$ and $\bigwedge_{j<\omega} \phi_i(x; a^i_j)$ is inconsistent. By the previous proposition, we may assume that the sequences I_i are mutually indiscernible over A. Then for $i < \kappa$, I_i is not indiscernible over Ab, which contradicts the definition of dp-rank. ⊣

We will show in Theorem 5.49 that forking equals dividing over extension bases, therefore one can replace "does not divide" by "does not fork" in the previous proposition.

PROPOSITION 5.48. *Let A be an extension base and $(a_i : i < \omega)$ indiscernible and strict non-forking over A. Assume that the formula $\phi(x; a_0)$ divides over A, then the partial type $\{\phi(x; a_i) : i < \omega\}$ is inconsistent.*

We say the strict non-forking sequences *witness dividing*.

PROOF. Assume that $\{\phi(x; a_i) : i < \omega\}$ is consistent. Then we may increase the sequence to some A-indiscernible sequence $(a_i : i < |T|^+)$, strict non-forking over A such that there is $b \models \bigwedge_{i<|T|^+} \phi(x; a_i)$. Then tp$(b/Aa_i)$ divides over a_i for each i, and this contradicts Proposition 5.47. ⊣

THEOREM 5.49. *Let A be an extension base and $\phi(x; a)$ a formula, then $\phi(x; a)$ forks over A if and only if it divides over A.*

PROOF. If $\phi(x; a)$ divides over A, then it forks over A. Conversely, assume that $\phi(x; a)$ forks over A. Then there are formulas $\psi_i(x; c) \in L(\mathcal{U}), i < n$, such that $\phi(x; a) \vdash \bigvee_{i<n} \psi_i(x; c)$ and each $\psi_i(x; c)$ divides over A. Let $q \in S(\mathcal{U})$ be a strict non-forking extension of tp(ac/A) (using Proposition 5.44) and let $(a_i c_i : i < \omega)$ be a Morley sequence of q over A. If $\phi(x; a)$ does not divide over A, then there is $b \models \bigwedge_{i<\omega} \phi(x; a_i)$. Trimming the sequence if necessary, we may assume that $b \models \bigwedge_{i<\omega} \psi_0(x; c_i)$. This contradicts Proposition 5.48. ⊣

Lowness. Let A be any set of parameters and let $\phi(x; y)$ be a formula over A. Let k be the alternation number of ϕ^{opp} where $\phi^{opp}(y; x) = \phi(x; y)$. So k is the maximal integer for which we can find an A-indiscernible sequence

$(b_i : i < \omega)$ such that $\bigwedge_{i<k} \neg(\phi(x; b_i) \leftrightarrow \phi(x; b_{i+1}))$ is consistent. Set also $l = \lfloor k/2 \rfloor + 1$. Let $(b_i : i < \omega)$ be any A-indiscernible sequence and assume that $\bigwedge_{i<\lceil k/2 \rceil} \phi(x; b_i)$ is consistent. We claim that $\bigwedge_{i<\omega} \phi(x; b_i)$ is consistent.

To see this, increase the sequence to an A-indiscernible $(b_i : i < \omega(l + 1))$. By indiscernibility, $\bigwedge_{1 \leq i \leq l} \phi(x; b_{\omega \cdot i})$ is consistent and let a realize it. Then there is some $i < l + 1$ such that a realizes $\bigwedge_{j<\omega} \phi(x; b_{\omega \cdot i+j})$, otherwise the truth value of $\phi(a; b_i)$ would alternate too often on the sequence. Thus again by indiscernibility, $\bigwedge_{i<\omega} \phi(x; b_i)$ is consistent.

PROPOSITION 5.50. *Let A be an extension base and $\phi(x; y)$ a formula over A. Then the set of b's such that $\phi(x; b)$ forks over A is type-definable over A.*

PROOF. By the paragraph before the proof (and Theorem 5.49), we know that $\phi(x; b)$ forks over A, if and only if it divides over A, if and only if there is some A-indiscernible sequence $(b_i : i < \omega)$ such that:

- $b_0 = b$;
- $\bigwedge_{i<\lfloor \mathrm{alt}(\phi^{opp})/2 \rfloor + 1} \phi(x; b_i)$ is inconsistent.

This is easily seen to be a type-definable condition on b. ⊣

Theories satisfying the conclusion of the previous proposition are called *low* theories. Thus NIP theories are low.

References and related subjects

The definition and general properties of forking and dividing are due to Shelah. See for example his book [106]. Properties of forking specific to NIP theories presented in Section 5.2 are proved in Shelah [109] and Hrushovski and Pillay [61]. The proof that forking equals dividing is from Chernikov and Kaplan [25]. We follow closely the simplified account given by Adler in [1]. The definition and properties of strict non-forking appeared in Shelah [109] and is investigated in Kaplan and Usvyatsov [70].

More information on NTP_2-theories can be found in Chernikov [24], and Chernikov and Ben Yaacov [18].

As we have seen, non-forking independence does not satisfy nice properties such as symmetry and transitivity. Hence it is tempting to look for a substitute independence relation. In o-minimal theories, such a notion is given by dimensional independence: a and b are independent over A if the dimension of the locus of a is the same over A and over Ab. An abstract relation generalizing this—called thorn-independence—can be defined in the class of *rosy* theories. Those notions were introduced by Scanlon and first investigated in works of Onshuus and Ealy: [85], [37]. Hasson and Onshuus study stable types in [54] and used rosiness in [55] to investigate structures definable in o-minimal theories. Krupiński studied rosy NIP groups in [72] and [73], whereas the work [36] by Ealy, Krupiński and Pillay focusses on *fsg* rosy groups.

Chapter 6

FINITE COMBINATORICS

In this chapter we introduce tools from finite combinatorics and probability theory.

It was first observed by Laskowski in [74] that NIP formulas had an analogue in combinatorics under the name *finite VC-dimension*. More precisely, a formula $\phi(x; y)$ is NIP if and only if the class $\{\phi(M; b) : b \in M\}$ of subsets of $M^{|x|}$ has finite VC-dimension. This is a central notion in machine learning theory, as classes of finite VC-dimension coincide with *learnable classes*. See [40]. This theory builds on the theorem of Vapnik and Chervonenkis (from which the name *VC* originates) which states a uniform law of large numbers for such classes.

We give a proof of that theorem and of the so-called *(p,q)*-theorem of Alon-Kleitman and Matoušek. We will therefore change the framework in this chapter and work first in a purely combinatorial setting. We come back to first order structures at the end to present the proof of uniformity of honest definitions, where the (p, q)-theorem plays an essential role.

6.1. VC-dimension

Let X be a set (finite or infinite) and $\mathcal{S}$ a family of subsets of X. Such a pair $(X, \mathcal{S})$ is called a *set system*. For most purposes, we can forget about the base set X, or in other words, take X to be $\bigcup \mathcal{S}$.

Let $A \subseteq X$. We say that the family $\mathcal{S}$ *shatters* A if for every $A' \subseteq A$, there is a set S in $\mathcal{S}$ such that $S \cap A = A'$.

The family $\mathcal{S}$ has VC-dimension at most n (written $\mathrm{VC}(\mathcal{S}) \leq n$), if there is no $A \subseteq X$ of cardinality $n + 1$ such that $\mathcal{S}$ shatters A. We say that $\mathcal{S}$ is of VC-dimension n if it is of VC-dimension at most n and shatters some subset of size n.

If for each n we can find a subset of X of cardinality n shattered by $\mathcal{S}$, then we say that $\mathcal{S}$ has infinite VC-dimension (and write $\mathrm{VC}(\mathcal{S}) = \infty$).

For a more careful analysis, we define the *shatter function* $\pi_{\mathcal{S}}$ from $\mathbb{N}$ to $\mathbb{N}$ as follows: $\pi_{\mathcal{S}}(n)$ is the maximum over all $A \subseteq X$ of cardinality $\leq n$ of

$|\{C \cap A : C \in \mathcal{S}\}|$. Note that $\pi_{\mathcal{S}}(n)$ is bounded by 2^n, and we have:

$$\mathrm{VC}(\mathcal{S}) < n \iff \pi_{\mathcal{S}}(n) < 2^n.$$

Given a set system $(X, \mathcal{S})$, we define the *dual* set system as the set system $(X^*, \mathcal{S}^*)$, where $X^* = \mathcal{S}$ and $\mathcal{S}^* = \{\mathcal{S}_a : a \in X\}$ with $\mathcal{S}_a = \{S \in \mathcal{S} : a \in S\}$. We then define the dual VC-dimension of $\mathcal{S}$ (written $\mathrm{VC}^*(\mathcal{S})$) as the VC-dimension of $\mathcal{S}^*$. We also define the dual shatter function $\pi^*_{\mathcal{S}}$ as the shatter function of $\mathcal{S}^*$.

Example 6.1. Let $X = \mathbb{R}^2$ and $\mathcal{S}$ be the set of open half planes. Then $\pi^*_{\mathcal{S}}(n)$ is the maximal number of regions into which n lines can partition the plane. One checks by induction that it is equal to $\frac{n(n+1)}{2} + 1$.

Example 6.2. Let $\phi(x; y)$ be a formula and fix a model M. Then $\phi(x; y)$ is NIP if and only if the class $\mathcal{S}_\phi(M) = \{\phi(M; b) : b \in M\}$ is of finite VC-dimension. The VC-dimension of $\phi(x; y)$ as defined in Section 2.1 coincides with the VC-dimension of that class. In particular, it is independent of the choice of M. We similarly define π_ϕ and the dual objects. Note that the dual VC-dimension of $\phi(x; y)$ is the VC-dimension of $\phi^{opp}(y; x)$ where $\phi^{opp}(y; x) = \phi(x; y)$.

LEMMA 6.3. *We have* $\mathrm{VC}^*(\mathcal{S}) < 2^{\mathrm{VC}(\mathcal{S})+1}$ *and* $\mathrm{VC}(\mathcal{S}) < 2^{\mathrm{VC}^*(\mathcal{S})+1}$.

In particular, $\mathcal{S}$ has finite VC-dimension if and only if it has finite dual VC-dimension.

PROOF. Assume $\mathrm{VC}(\mathcal{S}) \geq 2^n$. Then there is some subset $A \subseteq X$ of size 2^n shattered by $\mathcal{S}$. Write $A = \{a_C : C \subseteq n\}$. For each $k < n$, let $S_k \in \mathcal{S}$ be such that $S_k \cap A = \{a_C : \{k\} \subseteq C \subseteq n\}$. Then easily, the family $\{S_k : k \leq n\}$ is shattered by $\mathcal{S}^*$. It follows that $\mathrm{VC}^*(\mathcal{S}) \geq n$. This proves the second inequality. The first one is proved similarly. ⊣

The following fundamental lemma states that the shatter function $\pi_{\mathcal{S}}(n)$ is either always equal to 2^n, or has polynomial growth.

LEMMA 6.4 (Sauer-Shelah lemma). *Let $\mathcal{S}$ be a class of VC-dimension at most k. Then, for $n \geq k$, we have $\pi_{\mathcal{S}}(n) \leq \sum_{i=0}^{k} \binom{n}{i}$.*

In particular $\pi_{\mathcal{S}}(n) = O(n^k)$.

PROOF. First notice that the bound is tight: take $\mathcal{S}$ to be the family of subsets of X of cardinality $\leq k$. Then $\mathcal{S}$ has VC-dimension exactly k and we see that its shatter function is equal to the bound in the statement of the lemma. The idea of the proof is to reduce the situation to this case by modifying the elements of $\mathcal{S}$, making them as small as possible without changing neither the cardinality of the family nor its VC-dimension.

Fix an integer $n \geq k$. If $\mathcal{S}$ contradicts the bound, then this is also true of some finite subfamily of $\mathcal{S}$, so without loss we may assume that $\mathcal{S}$ is finite. Similarly, we may assume that X is finite of cardinality n, say $X = \{x_1, \ldots, x_n\}$, so that $\pi_{\mathcal{S}}(n) = |\mathcal{S}|$.

We define iteratively families $\mathcal{S}_0, \ldots, \mathcal{S}_n$. Set $\mathcal{S}_0 = \mathcal{S}$.

Let $l < n$ and assume that $\mathcal{S}_l$ has been defined. Go through the sets in $\mathcal{S}_l$ one by one. For each $S \in \mathcal{S}_l$, if $x_{l+1} \in S$, and if $S \setminus \{x_{l+1}\}$ is not a set in $\mathcal{S}_l$, replace S by $S \setminus \{x_{l+1}\}$. If not, leave S as it is. Let $\mathcal{S}_{l+1}$ be the resulting family.

The following three facts can be easily checked by induction on l:

(i) for each l, the cardinality of $\mathcal{S}_l$ is the same as that of $\mathcal{S}$;
(ii) let $S \in \mathcal{S}_l$ and $A = S \cap \{x_1, \ldots, x_l\}$; then for every $A_0 \subseteq A$, the set $A_0 \cup (S \setminus A)$ is in $\mathcal{S}_l$;
(iii) any $A \subseteq X$ shattered by $\mathcal{S}_{l+1}$ is also shattered by $\mathcal{S}_l$.

Fact (ii) implies in particular that if $S \in \mathcal{S}_n$, then S is shattered by $\mathcal{S}_n$. It follows from (iii) that the VC-dimension of $\mathcal{S}_n$ is not greater than that of $\mathcal{S}$. Therefore no set in $\mathcal{S}_n$ can have cardinality greater than k. Hence $\sum_{i=0}^{k} \binom{n}{i} \geq |\mathcal{S}_n| = |\mathcal{S}| = \pi_{\mathcal{S}}(n)$. ⊣

We define the *VC-density* of $\mathcal{S}$ to be $\mathrm{vc}(\mathcal{S}) = \limsup_{n\to\infty} \frac{\log(\pi_{\mathcal{S}}(n))}{\log n}$. In other words, $\mathrm{vc}(\mathcal{S})$ is the smallest $r \geq 0$ for which we have $\pi_{\mathcal{S}}(n) = O(n^r)$. Similarly, we define the *VC-codensity* as $\mathrm{vc}^*(\mathcal{S}) = \mathrm{vc}(\mathcal{S}^*)$.

We have $\mathrm{vc}(\mathcal{S}) < +\infty \iff \mathrm{VC}(\mathcal{S}) < +\infty$ and by the previous result, we always have $\mathrm{vc}(\mathcal{S}) \leq \mathrm{VC}(\mathcal{S})$. However, one cannot bound in general the VC-dimension in terms of the VC-density. For many purposes, the VC-density is a more appropriate notion than the VC-dimension.

Example 6.5. Let $(X, \mathcal{S})$ be a set system with $\mathrm{vc}(\mathcal{S}) < k$. Set $X' = X \cup Y$ where Y is a set of size k disjoint from X and let $\mathcal{S}' = \mathcal{S} \cup \mathfrak{P}(Y)$. Then $\mathrm{VC}(\mathcal{S}') = k$, but $\mathrm{vc}(\mathcal{S}') = \mathrm{vc}(\mathcal{S}) < k$.

We now state and prove the fundamental theorem of Vapnik and Chervonenkis. It is a uniform version of the law of large numbers for set systems of finite VC-dimension.

First, we fix a notation. For $S \in \mathcal{S}$ and $(x_1, \ldots, x_n) \in X^n$, we define $\mathrm{Av}(x_1, \ldots, x_n; S)$ as being equal to $\frac{1}{n}|S \cap \{x_1, \ldots, x_n\}|$. It is the measure of S estimated on the finite set $\{x_1, \ldots, x_n\}$. The weak law of large numbers (Proposition B.4) states that for fixed $S \in \mathcal{S}$ and $\varepsilon > 0$, we have $\mu^n(|\mathrm{Av}(x_1, \ldots, x_n; S) - \mu(S)| \geq \varepsilon) \leq \frac{1}{4n\varepsilon^2}$. Hence with high probability, sampling on a tuple $(x_1, \ldots, x_n)$ selected at random gives a good estimate of the measure of S. The VC-theorem states that if $\mathcal{S}$ is of finite VC-dimension, then sampling on a random tuple $(x_1, \ldots, x_n)$ gives a good estimate of the measures of all the sets in $\mathcal{S}$.

THEOREM 6.6 (VC-theorem). *Let (X, μ) be a finite probability space, and $\mathcal{S} \subseteq \mathfrak{P}(X)$ a family of subsets, then for $\varepsilon > 0$ we have*:

$$\mu^n\left(\sup_{S\in\mathcal{S}} |\mathrm{Av}(x_1, \ldots, x_n; S) - \mu(S)| > \varepsilon\right) \leq 8\pi_{\mathcal{S}}(n) \exp\left(-\frac{n\varepsilon^2}{32}\right).$$

PROOF. Fix some integer n. For $\bar{x} = (x_1, \ldots, x_n)$, $\bar{x}' = (x'_1, \ldots, x'_n)$ and $S \in \mathcal{S}$, let $f(\bar{x}, \bar{x}'; S)$ be equal to $|\operatorname{Av}(x_1, \ldots, x_n; S) - \operatorname{Av}(x'_1, \ldots, x'_n; S)|$.

Let $x_1, \ldots, x_n, x'_1, \ldots, x'_n$ be mutually independent random elements from X each with distribution μ. Let also $\sigma_1, \ldots, \sigma_n$ be random variables independent from each other and from the previous ones such that $\operatorname{Prob}(\sigma_i = 1) = \operatorname{Prob}(\sigma_i = -1) = 1/2$.

CLAIM 1. *We have*:

$$\operatorname{Prob}\left(\sup_{S\in\mathcal{S}} f(\bar{x}, \bar{x}'; S) > \varepsilon/2\right) \leq 2 \operatorname{Prob}\left(\sup_{S\in\mathcal{S}} \frac{1}{n}\left|\sum_i \sigma_i \boldsymbol{1}_S(x_i)\right| > \varepsilon/4\right).$$

PROOF OF CLAIM 1. Notice that for fixed i and S, the random variable $\boldsymbol{1}_S(x_i) - \boldsymbol{1}_S(x'_i)$ has zero mean and a symmetric distribution (it takes the values 1 and -1 with the same probability). Therefore its distribution does not change if we multiply it by σ_i. We then compute:

$$\begin{aligned}
&\operatorname{Prob}\left(\sup_{S\in\mathcal{S}} f(\bar{x}, \bar{x}'; S) > \varepsilon/2\right) \\
&\quad = \operatorname{Prob}\left(\sup_{S\in\mathcal{S}} \frac{1}{n}\left|\sum (\boldsymbol{1}_S(x_i) - \boldsymbol{1}_S(x'_i))\right| > \varepsilon/2\right) \\
&\quad = \operatorname{Prob}\left(\sup_{S\in\mathcal{S}} \frac{1}{n}\left|\sum \sigma_i(\boldsymbol{1}_S(x_i) - \boldsymbol{1}_S(x'_i))\right| > \varepsilon/2\right) \\
&\quad \leq \operatorname{Prob}\left(\sup_{S\in\mathcal{S}} \frac{1}{n}\left|\sum \sigma_i \boldsymbol{1}_S(x_i)\right| > \varepsilon/4 \text{ or } \sup_{S\in\mathcal{S}} \frac{1}{n}\left|\sum \sigma_i \boldsymbol{1}_S(x'_i)\right| > \varepsilon/4\right) \\
&\quad \leq 2 \operatorname{Prob}\left(\sup_{S\in\mathcal{S}} \frac{1}{n}\left|\sum \sigma_i \boldsymbol{1}_S(x_i)\right| > \varepsilon/4\right).
\end{aligned}$$

The last inequality being just the union bound. ⊣ (Claim 1)

CLAIM 2. *We have*:

$$\operatorname{Prob}\left(\sup_{S\in\mathcal{S}} f(\bar{x}, \bar{x}'; S) > \varepsilon/2\right) \leq 4\pi_{\mathcal{S}}(n) \exp\left(-\frac{n\varepsilon^2}{32}\right).$$

PROOF OF CLAIM 2. We start by fixing a tuple $\bar{a} \in X^n$ and some $S \in \mathcal{S}$. Let $A_S(\bar{a})$ be the event $\frac{1}{n}\left|\sum_i \sigma_i \boldsymbol{1}_S(a_i)\right| > \varepsilon/4$ (the only randomness left is in the σ_i's). By Chernoff's bound (Theorem B.5 and the remark following it) we have:

$$\operatorname{Prob}(A_S(\bar{a})) \leq 2 \exp\left(-\frac{n\varepsilon^2}{32}\right).$$

Now note that the event $A_S(\bar{a})$ depends only on $S \cap \{a_1, \ldots, a_n\}$. As S varies in $\mathcal{S}$, there are at most $\pi_{\mathcal{S}}(n)$ values for that set. Hence also at most $\pi_{\mathcal{S}}(n)$ events A_S to consider. Thus the union bound shows that the disjunction $\bigcup_{S\in\mathcal{S}} A_S(\bar{a})$

has probability at most $2\pi_{\mathcal{S}}(n)\exp(-n\varepsilon^2/32)$. By Claim 1, we have

$$\begin{aligned}\operatorname{Prob}\left(\sup_{S\in\mathcal{S}} f(\bar{x},\bar{x}';S) > \varepsilon/2\right) &\leq 2\operatorname{Prob}\left(\bigcup_{S\in\mathcal{S}} A_S(\bar{x})\right)\\ &\leq 4\pi_{\mathcal{S}}(n)\exp\left(-\frac{n\varepsilon^2}{32}\right). \quad \dashv \text{(Claim 2)}\end{aligned}$$

To conclude with the proof of the theorem, we may assume that $n > 2/\varepsilon^2$, since otherwise the right hand side is larger than 1. Let $X_0 \subseteq X^n$ be the set of $\bar{b} \in X^n$ such that $\operatorname{Prob}\left(\sup_{S\in\mathcal{S}} f(\bar{x},\bar{b};S) > \varepsilon/2\right) \geq 1/2$. By Claim 2, we have $\mu^n(X_0) \leq 8\pi_{\mathcal{S}}(n)\exp\left(-n\varepsilon^2/32\right)$. Fix $\bar{a} \in X^n \setminus X_0$ and some $S \in \mathcal{S}$. By the weak law of large numbers (Proposition B.4), we have

$$\operatorname{Prob}\left(|\operatorname{Av}(x_1,\dots,x_n;S) - \mu(S)| > \frac{\varepsilon}{2}\right) \leq \frac{1}{n\varepsilon^2} < \frac{1}{2}.$$

It follows that there is $\bar{x} \in X^n$ satisfying both:

- $f(\bar{x},\bar{a};S) \leq \varepsilon/2$;
- $|\operatorname{Av}(\bar{x};S) - \mu(S)| \leq \varepsilon/2$.

Remembering the definition of f, this implies that $|\operatorname{Av}(\bar{a};S) - \mu(S)| \leq \varepsilon$. As S was arbitrary, we conclude that for any $\bar{a} \in X^n \setminus X_0$, we have $\sup_{S\in\mathcal{S}} |\operatorname{Av}(\bar{a};S) - \mu(S)| \leq \varepsilon$ and the theorem follows. $\dashv$

In fact the theorem is still true if X is infinite, but we have to add the following measurability assumptions:

(i) each set $S \in \mathcal{S}$ is measurable;
(ii) for each n, the function

$$(x_1,\dots,x_n) \mapsto \sup_{S\in\mathcal{S}} |\operatorname{Av}(x_1,\dots,x_n;S) - \mu(S)|$$

from X^n to $\mathbb{R}$ is measurable;
(iii) for each n, the function

$$(x_1,\dots,x_n,x_1',\dots,x_n') \mapsto \sup_{S\in\mathcal{S}} |\operatorname{Av}(x_1,\dots,x_n;S) - \operatorname{Av}(x_1',\dots,x_n';S)|$$

from X^{2n} to $\mathbb{R}$ is measurable.

The first condition implies the other two when the family $\mathcal{S}$ is countable, and of course they always hold when X is finite. The proof then goes through unchanged.

Example 6.7. To see how the second and third hypothesis might fail, consider the case of $X = \omega_1$. Let $\mathcal{B}$ be the σ-algebra generated by the intervals. Let μ be defined on $\mathcal{B}$ by $\mu(A) = 1$ if A contains an end segment of X and $\mu(A) = 0$ otherwise. This defines a σ-additive measure on $(X, \mathcal{B})$. Take $\mathcal{S}$ to be the family of intervals of X. It has VC-dimension 2. We leave it to the reader to check that the VC-theorem does not hold for $\mathcal{S}$. (In view of Corollary 6.9 below it is enough to check that there are no ε-approximations, for $\varepsilon < 1$.)

DEFINITION 6.8. If $(X, \mathcal{S})$ is a set system, an ε-approximation for $\mathcal{S}$ is a finite subset $X_0 = \{x_1, \dots, x_n\}$ of X such that for all $S \in \mathcal{S}$, we have

$$\left| \frac{1}{n}|S \cap \{x_1, \dots, x_n\}| - \mu(S) \right| \leq \varepsilon.$$

COROLLARY 6.9. *Let $k > 0$ and $\varepsilon > 0$, then there is N such that any set system $\mathcal{S}$ on a finite probability space (X, μ) with $VC(\mathcal{S}) \leq k$ admits an ε-approximation of size at most N.*

PROOF. By Sauer's lemma 6.4, we know that there is some polynomial P_k depending only on k such that $\pi_{\mathcal{S}}(n) \leq P_k(n)$. Take N such that $8P_k(N) \exp\left(-\frac{N\varepsilon^2}{32}\right) < 1$ and apply the VC-theorem 6.6. ⊣

6.2. The (p, q)-theorem

Let $p \geq q$ be two integers. A set system $(X, \mathcal{S})$ has the (p, q)-property if out of every p sets of $\mathcal{S}$, some q have non-empty intersection. Dually, we will say that $(X, \mathcal{S})$ has the $(p, q)^*$-property if for every $X_1 \subseteq X$ of cardinality $\geq p$, there is $S \in \mathcal{S}$ such that $|X_1 \cap S| \geq q$.

The following theorem will be an important ingredient in the proof of uniformity for honest definitions.

THEOREM 6.10 ((p, q)-theorem). *Let $p \geq q$ be two integers. Then there is an integer N such that the following holds:*

Let $(X, \mathcal{S})$ be a finite set system where every $S \in \mathcal{S}$ is non-empty. Assume:

- *$VC^*(\mathcal{S}) < q$;*
- *$(X, \mathcal{S})$ has the (p, q)-property.*

Then there is a subset of X of size N which intersects every element of $\mathcal{S}$.

This theorem has the following model-theoretic consequence.

COROLLARY 6.11. *(T is NIP.) Fix a model M of T and let $\phi(x; y)$, $\psi(y)$ be two formulas such that $\phi(x; b)$ is non-forking over M for all $b \in \psi(\mathcal{U})$. Then there are finitely many global types $p_1, \dots, p_N \in S_x(\mathcal{U})$ such that any $\phi(x; b)$, $b \in \psi(\mathcal{U})$ is in one of them.*

PROOF. By assumption, one cannot find an indiscernible sequence $(b_i : i < \omega)$ of realizations of $\psi(y)$ such that $\{\phi(x; b_i) : i < \omega\}$ is inconsistent. Hence by compactness, for any q, there is some p such that for no subset $B_0 \subset \psi(\mathcal{U})$ of size p is $\{\phi(x; b) : b \in B_0\}$ q-inconsistent. This exactly means that the family $\{\phi(M; b) : b \in \psi(\mathcal{U})\}$ has the (p, q)-property. Take q large enough so that Theorem 6.10 applies and let N be given by it. Consider the partial type

$$q(x_0, \dots, x_{N-1}) = \left\{ \bigvee_{i<N} \phi(x_i; b) : b \in \psi(\mathcal{U}) \right\}.$$

By the (p, q)-theorem, every finite subset of $q(x_0, \dots, x_{N-1})$ is consistent. Hence the whole type is consistent and we obtain what we want. ⊣

Exercise 6.12. Show that in the previous corollary, one can ask for the types $p_1, \dots, p_N$ to be M-invariant.

We will only need a special case of the (p, q)-theorem where $p = q$ big enough with respect to $VC^*(\mathcal{S})$. We state this as a separate corollary, which we phrase in the dual form for later reference. We will only prove this corollary (without using the theorem) and refer the reader to [82] for the proof of the full theorem.

COROLLARY 6.13. *Let $k \in \mathbb{N}$, then there are two integers q and N such that for every finite X and $\mathcal{S} \subseteq \mathfrak{P}(X)$ a family of VC-dimension at most k, if $(X, \mathcal{S})$ has the $(q, q)^*$-property (that is, for every $X_0 \subseteq X$ of size $\leq q$, we can find $S \in \mathcal{S}$ containing X_0), then there are $S_1, \dots, S_N \in \mathcal{S}$ whose union is the whole of X.*

PROOF. (Of the corollary)

Let $\varepsilon = 1/3$.

By Corollary 6.9, there is some q such that for every set system $\mathcal{S}$ of VC-dimension $\leq k$ on a finite set X and any probability measure μ on X, there are $x_1, \dots, x_q \in X$ such that for any $S \in \mathcal{S}$,

$$\left| \mu(S) - \frac{|\{i : x_i \in S\}|}{n} \right| \leq \varepsilon.$$

Let $(X, \mathcal{S})$ be a set system of VC-dimension $\leq k$ with X finite and having the $(q, q)^*$-property. Then it follows that for any probability measure μ on X, we can find some $S \in \mathcal{S}$ with $\mu(S) \geq 1 - \varepsilon$.

We now need the following result, known as Farkas's lemma. We refer the reader to any introductory text on convex analysis or linear programing for a proof. See for example [100, page 90].

FACT 6.14 (Farkas's lemma). *Let A be a matrix in $\mathcal{M}_{m,n}(\mathbb{R})$ and $b \in \mathbb{R}^m$, then the following are equivalent:*

(i) *$\exists x \in \mathbb{R}^n_+$ such that $Ax \leq b$;*
(ii) *for all $y \in \mathcal{M}_{1,m}(\mathbb{R}_+)$, $yA \geq 0$ implies $yb \geq 0$.*

Write $X = \{x_1, \dots, x_{m-1}\}$ and $\mathcal{S} = \{S_1, \dots, S_n\}$. We define a matrix $A \in \mathcal{M}_{m,n}$ by:

- for all $j \leq n$, $A_{m,j} = 1$;
- for $i \leq m - 1$, $j \leq n$, $A_{i,j} = -1$ if $x_i \in S_j$ and 0 otherwise.

We also define $b \in \mathbb{R}^m$ by $b_i = -(1-\varepsilon)$ for $i \leq m-1$ and $b_m = 1$.

We check that condition (ii) of Farkas's lemma is satisfied. So let $y = (a_1, \dots, a_{m-1}, a_m) \in \mathcal{M}_{1,m}(\mathbb{R}_+)$ be such that $yA \geq 0$. This means that for any $j \leq n$, we have $\sum_{i=1}^m a_i A_{ij} \geq 0$. Hence, for any $j \leq n$, $\sum_{\{i: x_i \in S_j\}} a_i \leq a_m$.

Let $a_* = \sum_{i \leq m-1} a_i$. Let $\tilde{\mu}$ be the measure on X defined by giving weight a_i/a_* to the point x_i, for all $i \leq m-1$. Then the previous inequality implies that no set S_j has measure greater than a_m/a_*. On the other hand, by what we have shown above, there is some $S \in \mathcal{S}$ such that $\tilde{\mu}(S) \geq (1-\varepsilon)$. It follows that $a_m/a_* \geq 1-\varepsilon$. And this gives exactly $yb \geq 0$.

By Farkas's lemma, we conclude that (i) above holds, namely that there is some $d = (d_1, \dots, d_n)^{\mathrm{T}}$, with each $d_i \geq 0$ such that $Ad \leq b$. Decoding, this gives that for $i \leq m-1$, $\sum_{\{j: x_i \in S_j\}} d_j \geq 1-\varepsilon$ and $\sum_{j=1}^m d_j \leq 1$. Increasing d_1 if necessary, we may assume that actually $\sum_{j=1}^m d_j = 1$.

We now consider the dual set system: to $x \in X$, we associate $E_x = \{S \in \mathcal{S} : x \in S\}$. Let $\mathcal{E} = \{E_x : x \in X\}$. Then $\mathrm{VC}(\mathcal{E}) \leq 2^{k+1}$. Equip the set $\mathcal{S}$ with the measure μ^* defined by the weights d_i. By Corollary 6.9 again, we can find N depending only on k and ε, and $S'_1, \dots, S'_N \in \mathcal{S}$ such that for all $x \in X$,

$$\left| \mu^*(E_x) - \frac{|\{l : x \in S'_l\}|}{N} \right| \leq \varepsilon.$$

For any $x \in X$, we have $\mu^*(E_x) = \sum_{\{j: x \in S_j\}} d_j \geq 1-\varepsilon > \varepsilon$, thus $|\{l : x \in S'_l\}| > 0$. It follows that $\bigcup_{l=1}^N S'_l = X$. ⊣

6.3. Uniformity of honest definitions

We come back to the model theoretic context. We assume that our ambient theory T is NIP.

Recall from Theorem 3.13 (honest definitions) and the remark following it that given $M \models T$, $A \subseteq M$ and $\phi(x;b) \in L(M)$, there is some $\psi(x;z) \in L$ such that for any finite $A_0 \subseteq \phi(A;b)$ we can find $d \in A$ with $A_0 \subseteq \psi(A;d) \subseteq \phi(A;b)$.

We now address the question of uniformity of ψ with respect to ϕ. First, compactness gives a weak uniformity statement.

LEMMA 6.15. *Let $\phi(x;y) \in L$. For any formula $\psi(x;z)$ (where x is the same variable as in ϕ and z may vary), let an integer q_ψ be given. Then there are finitely many formulas $\psi_0, \dots, \psi_{n-1}$ such that:*

For $M \models T$, $A \subseteq M$, $b \in M$, there exists $j < n$, such that for any $A_0 \subseteq \phi(A;b)$ of size $\leq q_{\psi_j}$ there is some $d \in A$ with $A_0 \subseteq \psi_j(A;d) \subseteq \phi(A;b)$.

PROOF. Consider the language $L' = L \cup \{\boldsymbol{P}(x), c_b\}$, where c_b is a new constant. Consider the L'-theory T' axiomatized by T along with the sentences

Θ_ψ, for $\psi(x; z) \in L$, where

$$\Theta_\psi = \exists x_0, \dots, x_{q_\psi - 1} \in \boldsymbol{P}$$
$$\left(\bigwedge_{i<q_\psi} \phi(x_i; c_b) \wedge \neg \exists d \in \boldsymbol{P} \left(\bigwedge_{i<q_\psi} \psi(x_i; d) \wedge \forall x \in \boldsymbol{P} \left(\psi(x; d) \to \phi(x; c_b) \right) \right) \right).$$

By Theorem 3.13 as recalled above, the theory T' is inconsistent. Therefore by compactness there are finitely many formulas $\psi_0, \dots, \psi_{n-1}$ such that $T \cup \{\Theta_{\psi_0}, \dots, \Theta_{\psi_{n-1}}\}$ is inconsistent. This gives what we want. ⊣

THEOREM 6.16. *Let $\phi(x; y) \in L$. There exists $\psi(x; z) \in L$ such that for any $M \models T$, $A \subseteq M$ of size ≥ 2, $b \in M$ and $A_0 \subseteq \phi(A; b)$ finite, there is $d \in A$ with*

$$A_0 \subseteq \psi(A; d) \subseteq \phi(A; b).$$

PROOF. For any formula $\psi(x; z)$, let k_ψ be the VC-dimension of ψ and let $(q_\psi, N(\psi))$ be given by Corollary 6.13 for $k = k_\psi$. Apply the previous lemma with those q_ψ's. It gives us formulas $\psi_0, \dots, \psi_{n-1}$. For $i < n$, let

$$\Psi_i(x; z_1, \dots, z_{N(\psi_i)}) = \bigvee_{j=1}^{N(\psi_i)} \psi_i(x; z_j).$$

Let now $M \models T$, $A \subseteq M$ and $b \in M$. By the previous lemma, there is some $i < n$ such that for any $A_1 \subseteq \phi(A; b)$ of size $\leq q_{\psi_i}$, we can find $d \in A$ with $A_1 \subseteq \psi_i(A; d) \subseteq \phi(A; b)$.

Let $A_0 \subseteq \phi(A; b)$ be finite. Let

$$\mathcal{S} = \{\psi_i(A_0; d) : d \in A \text{ and } \psi_i(A; d) \subseteq \phi(A; b)\} \subseteq \mathfrak{P}(A_0).$$

Then the VC-dimension of $\mathcal{S}$ is bounded above by $\mathrm{VC}(\psi_i)$. Furthermore, the assumptions on ψ_i imply that $\mathcal{S}$ has the $(q_{\psi_i}, q_{\psi_i})^*$-property. Therefore by Corollary 6.13, there are $S_1, \dots, S_{N(\psi_i)}$ which cover the whole of A_0. Write S_j as $\psi_i(A_0; d_j)$. Then

$$A_0 \subseteq \Psi_i(A; d_1, \dots, d_{N(\psi_i)}) \subseteq \phi(A; b).$$

If $|A| \geq 2$, then by usual coding tricks, we can replace the finite set $\{\Psi_0, \dots, \Psi_{n-1}\}$ by a single formula and the theorem follows. ⊣

The following corollary is usually referred to as "uniform definability of types over finite sets", or UDTFS.

COROLLARY 6.17 (UDTFS). *Let $\phi(x; y) \in L$, then there is $\psi(x; z) \in L$ such that for any $b \in \mathcal{U}$ and $A \subset \mathcal{U}$ a finite set of size ≥ 2, there is $d \in A$ with*

$$\phi(A; b) = \psi(A; d).$$

PROOF. Simply apply Theorem 6.16 with $A_0 = \phi(A; b)$. ⊣

Note that this implies that the number of ϕ-types over A is bounded by $|A|^{|z|}$. One can thus see UDTFS as a model theoretic version of Sauer's lemma 6.4 which says that the number of ϕ-types is polynomial in the size of A. However we do not have any explicit bound on $|z|$.

Remark 6.18. We have used the fact that the theory T is NIP to obtain an honest definition for $\phi(x; b)$ and again to apply Corollary 6.13 to the formulas $\psi(x; z)$. In particular, the proof does not go through for an NIP formula $\phi(x; y)$ in a possibly independent theory. It is an open question whether or not UDTFS holds in that case.

Exercise 6.19 (Pseudofinite). Recall that is a theory T is pseudofinite if it is elementarily equivalent to an ultraproduct of finite structures.

We have seen that over finite sets, the number of types in a fixed number of variables n is polynomial in the size of the finite set. Furthermore, types over finite set are uniformly definable. Hence it appears that over finite sets, NIP theories behave like stable theories. It is then reasonable to expect that pseudofinite NIP theories will share other properties with stable theories.

1. Characterize pseudofinite linear orders.
   Assume that T is NIP and pseudofinite.
2. Show that any definable set in T is stably embedded.
3. (Chernikov) Show that if T is moreover ω-categorical, then it is stable. (It is an open question whether we can replace ω-categorical by elimination of $\exists^\infty$.)
4. Show that if some formula $\phi(x; a)$ does not fork over a model M, then it has a point in M.

References and related subjects

VC-dimension was introduced in the context of statistical learning theory by Vapnik and Chervonenkis in [120] where Theorem 6.6 is proved. It has played a major role in that area. Sauer's lemma 6.4 is implicit in [120], and was rediscovered independently by at least two authors: Shelah [104] and Sauer [98]. Shelah and Sauer's proofs are by induction on n. The proof we give was found independently by Alon [5] and Frankl [41].

The (p, q) theorem was proved for families of convex subsets of the euclidian space by Alon and Kleitman [6] and then for families of finite VC-dimension by Matoušek [82]. The UDTFS property was conjectured by Laskowski and proved for o-minimal theories by him and Johnson in [64]. Then Guingona proved it for dp-minimal theories [47]. It was proved for all NIP theories in Chernikov and Simon [27]. This problem is linked to a similar question in machine learning theory (existence of *compression schemes* for learnable families). See the discussion and references in Johnson and Laskowski [64] or in Livni and Simon [75]. The latter paper also gives an *effective version* of the proof of UDTFS.

Two papers [8] and [9] by Aschenbrenner, Dolich, Haskell, Macpherson and Starchenko undertake a quantitative study of *VC-density* of formulas in various NIP theories. In the first paper, they obtain bounds on VC-densities of formulas (as a function of the number of parameters) notably in o-minimal theories, in ACVF and the p-adics. They also give examples of formulas with non-integer VC-density or whose shatter function is not asymptotically equivalent to a power function. The second paper focusses on stable theories. The authors give explicit bounds for theories of finite U-rank without the finite cover property and characterize abelian groups for which there is a uniform bound on the VC-density of formulas.

Chapter 7

MEASURES

In this chapter we study Keisler measures, which can be seen either as a generalization of types that allow for truth values in the segment $[0, 1]$ or as ordinary probability measures on the compact space of types. The two points of view are useful and we will often switch from one to the other.

Many properties of types generalize naturally to measures. In particular one can define generically stable measures. As we will see, those are ubiquitous in NIP theories, as opposed to generically stable *types*. One way to obtain such a measure is to take a σ-additive probability measure on a standard model (for example the Lebesgue measure on $[0, 1]$ seen as subset of the structure $\mathbb{R}$).

The most important examples of measures are invariant measures on groups. Those will be studied in the next chapter.

In this chapter, we again assume that T is NIP, although the basic definitions are of course valid in any theory.

7.1. Definitions and basic properties

If A is a set of parameters and x a variable, we let $\mathcal{L}_x(A)$ denote the algebra of A-definable sets in the variable x. Equivalently, it is the Boolean algebra of formulas with parameters in A and free variable x, quotiented by the equivalence relation $\phi(x) \sim \psi(x) \iff \mathcal{U} \models \phi(x) \leftrightarrow \psi(x)$. By an abuse of notations, $\phi(x)$ will be used to denote a formula as well as its image in $\mathcal{L}_x(A)$.

Definition 7.1. Let $A \subset \mathcal{U}$ be a set of parameters. A *Keisler measure* (or simply a *measure*) μ over A in the variable x is a finitely additive probability measure on $\mathcal{L}_x(A)$. In other words it is a function $\mu : \mathcal{L}_x(A) \to [0, 1]$ such that:

- $\mu(x = x) = 1$;
- $\mu(\neg\phi(x)) = 1 - \mu(\phi(x))$;
- $\mu(\phi(x) \wedge \psi(x)) + \mu(\phi(x) \vee \psi(x)) = \mu(\phi(x)) + \mu(\psi(x))$.

We will sometimes write μ as μ_x or $\mu(x)$ to emphasize that μ is a measure on the variable x. If $A \subseteq B$ and μ is a measure over B, we define the restriction of μ to A denoted $\mu|_A$ or $\mu \upharpoonright A$ as the restriction of μ to $\mathcal{L}_x(A)$. As in the case of types, we say that μ is an extension to B of $\mu|_A$.

Example 7.2.
- A type $p \in S_x(A)$ can (and will) be identified with the Keisler measure $\mu_p(x)$ over A defined by $\mu_p(\phi(x)) = 1$ if $p \vdash \phi(x)$ and $\mu_p(\phi(x)) = 0$ otherwise. Thus a (complete) type is a special case of a Keisler measure. We will usually not distinguish between p and μ_p and write for example $p(\phi(x;b))$ instead of $\mu_p(\phi(x;b))$.
- Given $a_0, a_1, \dots$ in $[0,1]$ such that $\sum a_i = 1$, and types $p_0, p_1, \dots$ over A in the same variable x, we can define the average measure $\mu = \sum a_i p_i$.
- Take T to be the theory of real closed fields and let $\mathbb{R}$ be the standard model. Let μ_0 be any Borel probability measure on $\mathbb{R}$. Then μ_0 induces a Keisler measure over $\mathcal{U}$ in one variable x defined by $\mu(\phi(x)) = \mu_0(\phi(\mathbb{R}))$.
- Let $I = (a_i : i \in [0,1])$ be an indiscernible sequence. Let λ_0 denote the usual Lebesgue measure on the interval $[0,1]$. We can define the *average measure* $\mathrm{Av}(I)$ as the measure μ defined by $\mu(\phi(x;b)) = \lambda_0(\{i \in [0,1] : \models \phi(a_i;b)\})$. Note that NIP ensures that the set in question is Lebesgue measurable (it is in fact a finite union of intervals).
- Let $(M_n : n < \omega)$ be a sequence of finite structures. For each $n < \omega$, let μ_n denote the normalized counting measure on M_n. Let $\mathcal{D}$ be an ultrafilter on ω and consider the ultraproduct $M = \prod_{\mathcal{D}} M_n$. We define a measure μ on M in the following way. Let $\phi(x;b) \in L(M)$ be a formula. Let $(b_n : n < \omega)$ be a representative of b in $\prod_{n<\omega} M_n$. Then set $\mu(\phi(x;b)) = \lim_{\mathcal{D}} \mu_n(\phi(M_n;b_n))$[1].
- Here is an example with IP which does not have all the nice properties that we will establish for measure in NIP theories. Let T be the theory of the random graph in the language $\{R\}$. Let $M \models T$. Define a Keisler measure μ on M by $\mu\left(\bigcap_{i<n}(xRa_i)^{\eta(i)}\right) = 2^{-n}$ for any choice of pairwise distinct a_i's in M and $\eta : n \to \{0,1\}$.

Let $\mathfrak{M}_x(A)$ denote the set of Keisler measures over A. It is a closed subset of $[0,1]^{\mathcal{L}_x(A)}$, equipped with the product topology. We equip $\mathfrak{M}_x(A)$ with the induced topology, making it a compact Hausdorff space. The identification of a type with the measure it defines gives an identification of $S_x(A)$ as a closed subspace of $\mathfrak{M}_x(A)$.

Borel measures. Let $\mu \in \mathfrak{M}_x(A)$ be a Keisler measure. It assigns a measure to every clopen set of the space $S_x(A)$. We show how to extend that measure to a σ-additive Borel probability measure. First, if $O \subseteq S_x(A)$ is open, we define $\mu(O) = \sup\{\mu(D) : D \subseteq O, D \text{ clopen}\}$. Similarly, the measure of a closed set F is the infimum of the measures of clopen sets which contain it. If $F \subseteq O$ are

[1] We recall that the limit $\lim_{\mathcal{D}} a_i$ of a sequence $(a_i)_{i<\omega}$ of elements of a compact set C is the unique $l \in C$ such that for any neighborhood U of l, the set $\{i < \omega : a_i \in U\}$ is in $\mathcal{D}$.

respectively closed and open, then there is a definable set between them. This implies that if X is either closed or open, we have

$$(Reg) \quad \sup\{\mu(F) : F \subseteq X, F \text{ closed}\} = \inf\{\mu(O) : X \subseteq O, O \text{ open}\}.$$

The next step is to show that the set of subsets $X \subseteq S_x(A)$ satisfying (Reg) is closed under complement and countable union. Complement is clear. For countable union: let $X = \bigcup_{i<\omega} X_i$ and fix $\varepsilon > 0$. For each $i < \omega$, take $F_i \subseteq X_i \subseteq O_i$ with $\mu(O_i) - \mu(F_i) \leq \varepsilon 2^{-i}$. Let $O = \bigcup_{i<\omega} O_i$. Then we can find some finite N such that $\mu(O) - \mu(\bigcup_{i<N} O_i) \leq \varepsilon$. Let $F = \bigcup_{i<N} F_i$. Then we have $F \subseteq X \subseteq O$ and $\mu(O) - \mu(F) \leq 3\varepsilon$.

It follows that every Borel subset of $S_x(A)$ satisfies (Reg). We can therefore define μ on all such sets by $\mu(X) = \sup\{\mu(F) : F \subseteq X, F \text{ closed}\} = \inf\{\mu(O) : X \subseteq O, O \text{ open}\}$. It is easy to check that this defines a σ-additive measure on $S_x(A)$. Property (Reg) is referred to as *regularity* of the measure μ.

To a Keisler measure on A, we have associated a regular probability measure on $S_x(A)$. Conversely, if μ is a regular probability measure on $S_x(A)$, then it defines a Keisler measure by restriction to the clopens. Regularity ensures that μ is entirely determined by that restriction. We therefore obtain a bijection:

$$\{\text{Keisler measures on } A\} \longleftrightarrow \{\text{Regular Borel prob. measures on } S_x(A)\}.$$

We will always use the same notation for the Keisler measure and for the associated Borel measure on the type space. This gives meaning to the notation $\mu(X)$, where μ is a Keisler measure on A and X is a Borel subset of $S_x(A)$.

We define the *support* of μ as the set $S(\mu) \subseteq S_x(A)$ of types $p \in S_x(A)$ such that for any $\phi(x;b) \in L(A)$, if $p(\phi(x;b)) = 1$ then $\mu(\phi(x;b)) > 0$. The support of μ is thus a closed set of $S_x(A)$. A type in the support of μ is called *weakly random* for μ.

If X is a Borel set of positive μ-measure, the *localization* of μ at X is the measure $\mu_X(x)$ defined by $\mu_X(\phi(x)) = \mu(\phi(x) \cap X)/\mu(X)$.

Extending measures. The next lemma says that any partial measure in a suitable sense extends to a full Keisler measure. In the statement, $\top$ denotes the true formula $x = x$.

Lemma 7.3. *Let $\Omega \subseteq \mathcal{L}_x(A)$ be a set of (equivalence classes of) formulas closed under intersection, union and complement, and containing $\top$. Let μ_0 be a finitely additive measure on Ω with values in $[0,1]$ such that $\mu_0(\top) = 1$. Then μ extends to a Keisler measure over A.*

Proof. By compactness in the space $[0,1]^{\mathcal{L}_x(A)}$, it is enough to show that given $\psi_1(x), \ldots, \psi_n(x)$ definable sets in $\mathcal{L}_x(A)$, there is a function $f = \langle \psi_1, \ldots, \psi_n \rangle \to [0,1]$ finitely additive and compatible with μ_0. (Here $\langle X \rangle$ denotes the Boolean algebra generated by X.) We may assume that $\psi_1, \ldots, \psi_n$ are the atoms of the Boolean algebra B that they generate.

The elements of Ω in B form a sub-Boolean algebra. Let $\phi_1, \dots, \phi_m$ be its atoms. We have say:

$$\phi_1 = \psi_{i_1(1)} \vee \dots \vee \psi_{i_1(l_1)} \quad \phi_2 = \psi_{i_2(1)} \vee \dots \vee \psi_{i_2(l_2)} \quad \text{etc.}$$

No ψ_i appears in more than one such expression since the ϕ_i's are disjoint. Then any finitely additive f satisfying $f(\psi_{i_1(1)}) + \dots + f(\psi_{i_1(l_1)}) = \mu_0(\phi_1)$ etc. will do. $\dashv$

LEMMA 7.4. *Let $\mu \in \mathfrak{M}_x(M)$ be a measure and let $\phi(x;b) \in L(\mathcal{U})$. Let $r_1 = \sup\{\mu(\psi(x)) : \psi(x) \in L(M), \models \psi(x) \to \phi(x;b)\}$ and $r_2 = \inf\{\mu(\psi(x)) : \psi(x) \in L(M), \models \phi(x;b) \to \psi(x)\}$.*

Then for any $r_1 \leq r \leq r_2$, there is an extension $\nu \in \mathfrak{M}_x(\mathcal{U})$ of μ such that $\nu(\phi(x;b)) = r$.

PROOF. It is enough to find some ν_1, ν_2 such that $\nu_i(\phi(x;b)) = r_i$ for $i \in \{1, 2\}$, since we can then consider averages of ν_1 and ν_2. We build ν_2. Let $\Omega \subseteq \mathcal{L}_x(\mathcal{U})$ be the Boolean algebra generated by $\mathcal{L}_x(M)$ and the definable set $\phi(x;b)$. By Lemma 7.3 it is enough to define ν_2 on Ω.

First assume that $r_2 = 0$. Then for any $\theta(x) \in L(M)$, set $\nu_2(\theta(x)) = \mu(\theta(x))$ and $\nu_2(\phi(x;b)) = 0$. This extends uniquely to a finitely additive measure on Ω. A similar argument works for $r_2 = 1$.

If $r_2 > 0$, then let μ' be the localization of μ on the closed set $C = \bigwedge \theta(x)$ where $\theta(x)$ ranges over elements of $\mathcal{L}_x(M)$ such that $\models \phi(x;b) \to \theta(x)$. Let μ'' be the localization of μ on the complementary open set. Then we have $\mu = r_2\mu' + (1 - r_2)\mu''$. By the previous paragraph, we can extend μ' and μ'' respectively to ν' and ν'' such that $\nu'(\phi(x;b)) = 1$ and $\nu''(\phi(x;b)) = 0$. Then set $\nu_2 = r_2\nu' + (1 - r_2)\nu''$.

The construction of ν_1 follows by taking $\neg\phi(x;b)$ instead of $\phi(x;b)$. $\dashv$

When manipulating types, it is often convenient to be able to realize them as points in a structure. The same thing cannot be done so easily with measures, but there are a couple of techniques that can be used instead. One of them is smooth measures which we will describe later. Another one consists in adding the interval [0, 1] as a new sort and coding the measure inside the new structure via definable maps.

More precisely, let M be a structure and let $\mu(x)$ be a Keisler measure over M. We encode this data as a structure

$$\tilde{M}_\mu = (M, [0,1], <, +, f_\phi : \phi(x;y) \in L),$$

where M is equipped with its full structure, [0, 1] is the standard unit interval endowed with the ordering and addition modulo 1. For each formula $\phi(x;y) \in L$ the function $f_\phi(y)$ sends b to $\mu(\phi(x;b)) \in [0,1]$. Let now $\tilde{N} = (N, [0,1]^*, \dots)$ be an elementary extension of $\tilde{M}_\mu$, where $[0,1]^*$ a (possibly) non-standard interval. Let st : $[0,1]^* \to [0,1]$ be the standard part map, that is the map sending any element of $[0,1]^*$ to the unique real number infinitesimally close to it. From $\tilde{N}$ we recover a Keisler measure $\mu'(x)$ over N by setting $\mu'(\phi(x;b)) = \mathrm{st}(f_\phi(b))$.

7.2. Boundedness properties

LEMMA 7.5. *Let $\mu \in \mathfrak{M}_x(M)$ be a measure and $(b_i : i < \omega)$ an indiscernible sequence in M. Let $\phi(x;y)$ be a formula and $r > 0$ such that $\mu(\phi(x;b_i)) \geq r$ for all $i < \omega$. Then the partial type $\{\phi(x;b_i) : i < \omega\}$ is consistent.*

PROOF. First we show that we can assume that the sequence $(b_i)_{i<\omega}$ is μ-indiscernible, by which we mean that if $i_1 < \cdots < i_n < \omega$ and $j_1 < \cdots < j_n < \omega$, then $\mu(\phi(x;b_{i_1}) \wedge \cdots \wedge \phi(x;b_{i_n})) = \mu(\phi(x;b_{j_1}) \wedge \cdots \wedge \phi(x;b_{j_n}))$. To see this, expand M to $\tilde{M}_\mu = (M, [0,1], \dots)$ as described after Lemma 7.4. Then in an elementary extension of $\tilde{M}_\mu$, we can find a sequence $(b'_i)_{i<\omega}$, indiscernible over $\tilde{M}_\mu$ in the extended language and realizing the EM-type of $(b_i)_{i<\omega}$. As $(b_i)_{i<\omega}$ is L-indiscernible, the two sequences have the same L-type. Therefore we may replace $(b_i)_{i<\omega}$ by $(b'_i)_{i<\omega}$, which has the desired property.

Now assume for a contradiction that $\{\phi(x;b_i) : i < \omega\}$ is inconsistent. Then there is some N such that $\mu(\phi(x;b_0) \wedge \cdots \wedge \phi(x;b_{N-1})) = 0$. Take a minimal such N and let $N' = N - 1$. For any integer m, let $\psi_m(x) = \phi(x;b_{mN'}) \wedge \cdots \wedge \phi(x;b_{mN'+N'-1})$. By minimality of N and indiscernibility of the sequence, there is some $t > 0$ such that we have $\mu(\psi_m(x)) = t$ for all m. Also by the property of N, we have $\mu(\psi_m(x) \wedge \psi_{m'}(x)) = 0$ for $m \neq m'$.

But then we have $\mu(\psi_0(x) \vee \cdots \vee \psi_{m-1}(x)) = mt$, for all m. This contradicts the fact that the measure of the whole space is 1. ⊣

Until now we have not used the assumption that T is NIP. It comes into play in the following lemma which implies that measures are bounded objects.

LEMMA 7.6. *Let $\mu \in \mathfrak{M}_x(M)$. We cannot find a sequence $(b_i : i < \omega)$ of tuples of M, a formula $\phi(x;y)$ and $\varepsilon > 0$ such that $\mu(\phi(x;b_i) \triangle \phi(x;b_j)) > \varepsilon$ for all $i, j < \omega$, $i \neq j$.*

PROOF. Assume otherwise. Then by Ramsey and compactness, we may assume that the sequence $(b_i : i < \omega)$ is indiscernible. By NIP, the partial type $\{\phi(x;b_{2k}) \triangle \phi(x;b_{2k+1}) : k < \omega\}$ is inconsistent. This contradicts the previous lemma. ⊣

Let $\mu \in \mathfrak{M}_x(\mathcal{U})$ be a global measure. If $\phi(x)$ and $\psi(x)$ are two M-definable sets, set $\phi(x) \sim_\mu \psi(x)$ if $\mu(\phi(x) \triangle \psi(x)) = 0$. Then the previous lemma implies that the set of $\sim_\mu$-equivalence classes has small cardinality (bounded by some κ depending only on $|T|$). In particular, the support $S(\mu)$ of μ has small cardinality.

7.3. Smooth measures

We now define and investigate the notion of a smooth measure, which can be considered as an analog of a realized type.

DEFINITION 7.7. Let $\mu \in \mathfrak{M}_x(M)$. We say that μ is *smooth* if for every $N \supseteq M$, μ has a unique extension to an element of $\mathfrak{M}_x(N)$.

More generally, if $\mu \in \mathfrak{M}_x(N)$ and $M \subseteq N$, we say that μ is smooth over M if $\mu|_M$ is smooth.

LEMMA 7.8. *Let $\mu \in \mathfrak{M}_x(M)$ be a smooth measure. Let $\phi(x, y) \in L$ and $\varepsilon > 0$. Then there are formulas $\theta_i^1(x), \theta_i^2(x)$ for $i = 1, \dots, n$ and $\psi_i(y)$ for $i = 1, \dots, n$, all over M such that:*

(i) *the formulas $\psi_i(y)$ partition y-space;*
(ii) *for all i, if $\models \psi_i(b)$, then $\models \theta_i^1(x) \to \phi(x, b) \to \theta_i^2(x)$;*
(iii) *for each i, $\mu(\theta_i^2(x)) - \mu(\theta_i^1(x)) < \varepsilon$.*

PROOF. Let $b \in \mathcal{U}$. Then by smoothness of μ and Lemma 7.4, there are formulas $\theta_i^1(x)$, $\theta_i^2(x)$ over M, such that

$$\models \theta_i^1(x) \to \phi(x, b) \to \theta_i^2(x), \text{ and } \mu(\theta_i^2(x)) - \mu(\theta_i^1(x)) < \varepsilon. \qquad (*)$$

By compactness, there are finitely many such pairs, say, $(\theta_i^1(x), \theta_i^2(x))$ such that for every b, one of these pairs satisfies $(*)$. It is then easy to find the $\psi_i(y)$'s. ⊣

Note that conversely, the existence of formulas $\theta_i^t(x)$ and $\psi_i(y)$ with the above properties implies that the measure μ is smooth.

PROPOSITION 7.9. *Let $\mu \in \mathfrak{M}_x(M)$ be any measure. Then there is $M \prec N$ and an extension $\mu' \in \mathfrak{M}_x(N)$ of μ which is smooth.*

PROOF. Assume not. We build an increasing sequence $(M_\alpha : \alpha < |T|^+)$ of models and an associated increasing sequence of measures $\mu_\alpha \in \mathfrak{M}_x(M_\alpha)$. Set $M_0 = M$. At limit stages, take the union. Assume M_α, μ_α are defined. By hypothesis, μ_α is not smooth, so we can find some $\phi_\alpha(x; b_\alpha) \in L(\mathcal{U})$ and μ^1, μ^2 two extensions of μ_α such that $\mu^2(\phi_\alpha(x; b_\alpha)) - \mu^1(\phi_\alpha(x; b_\alpha)) = 4\varepsilon_\alpha > 0$. Then take $M_{\alpha+1}$ an extension of M_α containing b_α and set $\mu_{\alpha+1} = 1/2(\mu^1 + \mu^2)$. Observe that for any $\theta(x) \in M_\alpha$, either the μ^1 or the μ^2 measure of $\theta(x) \triangle \phi_\alpha(x; b_\alpha)$ is $\geq 2\varepsilon_\alpha$, hence $\mu_{\alpha+1}(\theta(x) \triangle \phi_\alpha(x; b_\alpha)) \geq \varepsilon_\alpha$.

Having done the construction, we may assume that $\phi_\alpha = \phi$, and $\varepsilon_\alpha = \varepsilon > 0$ are both constant. Let μ' be the measure $\bigcup_{\alpha < |T|^+} \mu_\alpha$. Then we have

$$\mu'(\phi(x; b_\alpha) \triangle \phi(x; b_\beta)) \geq \varepsilon$$

for any $\alpha < \beta$. This contradicts Lemma 7.6. ⊣

We now show that smooth measures in NIP theories can be approximated by averages of points.

If $\mu_0, \dots, \mu_{n-1}$ are measures and $\phi(x;b)$ is a formula, then the notation $\mathrm{Av}(\mu_0, \dots, \mu_{n-1}; \phi(x;b))$ stands for $\frac{1}{n}\sum_{k<n} \mu_k(\phi(x;b))$.

Similarly, if $a_0, \dots, a_{n-1}$ are tuples, then $\mathrm{Av}(a_0, \dots, a_{n-1}; \phi(x;b))$ stands for $\frac{1}{n}|\{i : \models \phi(a_i;b)\}|$. We extend this notation naturally to the case where $\phi(x;b)$ is replaced by an arbitrary Borel subset of some type space.

PROPOSITION 7.10. *Let $\mu(x)$ be a global measure, smooth over M. Let X be a Borel subset of $S_x(M)$, and $\phi(x;y)$ a formula. Fix $\varepsilon > 0$. Then there are $a_0, \dots, a_{n-1} \in \mathcal{U}$ such that for any $b \in \mathcal{U}$,*

$$|\mu(X \cap \phi(x;b)) - \mathrm{Av}(a_0, \dots, a_{n-1}; X \cap \phi(x;b))| \leq \varepsilon.$$

PROOF. Fix formulas $\psi_i(y)$, $\theta_i^0(x)$ and $\theta_i^1(x)$, $i < m$ as given by Lemma 7.8. Consider μ as a probability measure on the space $S_x(M)$. We can find types $(p_i : i < n)$ in $S_x(M)$ such that if $\lambda = \frac{1}{n}\sum_{i<n} p_i$ then for all $i < m$, $\lambda(\theta_i^0(x) \cap X)$ and $\lambda(\theta_i^1(x) \cap X)$ are within ε of $\mu(\theta_i^0(x) \cap X)$ and $\mu(\theta_i^1(x) \cap X)$ respectively. (For example, pick types p_i at random according to μ and apply the weak law of large numbers.)

Now take points $(a_i : i < n)$ in $\mathcal{U}$ such that $a_i \models p_i$. Set $\lambda' = \frac{1}{n}\sum_{i<n} \mathrm{tp}(a_i/\mathcal{U})$. Let $b \in \mathcal{U}$ and let $i < n$ be such that $\models \psi_i(b)$. Then we have $\models \theta_i^0(x) \to \phi(x;b) \to \theta_i^1(x)$ and $\mu(\theta_i^1(x)) - \mu(\theta_i^0(x)) \leq \varepsilon$. This implies that $\lambda(\theta_i^1(x) \cap X) - \lambda(\theta_i^0(x) \cap X) \leq 3\varepsilon$ and therefore $\lambda'(\phi(x;b) \cap X)$ is within 3ε of $\lambda(\phi(x;b) \cap X)$ and thus within 4ε of $\mu(\phi(x;b) \cap X)$. ⊣

We conclude that we can approximate any measure by averages of types.

PROPOSITION 7.11. *Let $\mu \in \mathfrak{M}_x(A)$ be any Keisler measure; let $\phi(x;y) \in L$ be a formula and fix $X_1, \dots, X_m \subseteq S_x(A)$ Borel subsets. Let $\varepsilon > 0$. Then there are types $p_0, \dots, p_{n-1} \in S_x(A)$ such that, for every $b \in A$ and every $k \leq m$:*

$$|\mu(\phi(x;b) \cap X_k) - \mathrm{Av}(p_0, \dots, p_{n-1}; \phi(x;b) \cap X_k)| \leq \varepsilon.$$

Furthermore, one can impose that $p_k \in S(\mu)$ for all k.

PROOF. Let ν_x be a smooth extension of μ over a model $N \supset A$. The previous proposition works equally well with finitely many Borel subsets instead of one and gives points $a_0, \dots, a_{n-1} \in N$ such that

$$|\nu(\phi(x;b) \cap X_k) - \mathrm{Av}(a_0, \dots, a_{n-1}; \phi(x;b) \cap X_k)| \leq \varepsilon$$

for all $b \in N$ and $k \leq m$. Then set $p_i = \mathrm{tp}(a_i/A)$.

To prove the "furthermore" part, apply the theorem with $S(\mu)$ as one of the Borel sets and assume without loss that one instance of $\phi(x;b)$ is equivalent to $x = x$. Then at most an ε fraction of the p_i's are not in $S(\mu)$. We can safely remove them: this adds at most an extra error of ε to the approximation. ⊣

Exercise 7.12. Give another proof of Proposition 7.11 using the VC-theorem instead of smooth measures. Deduce that the number n of types can be chosen so as to depend only on VC-dim$(\phi(x;y))$ and ε (see [61, Lemma 4.8]).

7.4. Invariant measures

We now extend a number of definitions from types to measures.

DEFINITION 7.13. Let $\mu \in \mathfrak{M}_x(\mathcal{U})$ be a measure, and let $A \subset \mathcal{U}$. We say that μ is A-invariant if for every $b \equiv_A b'$, and $\phi(x;y) \in L$, we have $\mu(\phi(x;b)) = \mu(\phi(x;b'))$.

We define Lstp(A)-invariant measures similarly.

DEFINITION 7.14. Let $A \subseteq B$ and $\mu \in \mathfrak{M}_x(B)$. Then μ *does not fork* (resp. divide) over A if $\mu(\phi(x;b)) = 0$ for every formula $\phi(x;b) \in L(B)$ which forks (resp. divides) over A.

PROPOSITION 7.15. *Let $\mu \in \mathfrak{M}_x(\mathcal{U})$ and $A \subset \mathcal{U}$. Then μ does not fork over A if and only if it is* Lstp(A)*-invariant.*

PROOF. Let $\mu \in \mathfrak{M}_x(\mathcal{U})$ be Lstp(A)-invariant. As it is a global measure, it is enough to show that it does not divide over A. Let $\phi(x;b) \in L(M)$ be such that $\mu(\phi(x;b)) > 0$. Let $(b_i : i < \omega)$ be an A-indiscernible sequence with $b_0 = b$. Then by assumption $\mu(\phi(x;b_i)) = \mu(\phi(x;b))$ for all $i < \omega$. By Lemma 7.5, the partial type $\{\phi(x;b_i) : i < \omega\}$ is consistent. It follows that $\phi(x;b)$ does not divide over A.

Conversely, assume that μ does not fork over A. Then if Lstp$(b/A) =$ Lstp(b'/A), the formula $\phi(x;b)\triangle\phi(x;b')$ forks over A. We conclude that $\mu(\phi(x;b)\triangle\phi(x;b')) = 0$. ⊣

DEFINITION 7.16. Let $M \models T$ and let $\mu \in \mathfrak{M}_x(\mathcal{U})$:

- μ is *finitely satisfiable* in M if for every $\phi(x;b) \in L(\mathcal{U})$ such that $\mu(\phi(x;b)) > 0$, there is $a \in M$ such that $\mathcal{U} \models \phi(a;b)$.
- μ is *definable* over M if it is M-invariant and for every $\phi(x;y) \in L$, and $r \in [0,1]$, the set $\{q \in S_y(M) : \mu(\phi(x;b)) < r \text{ for any } b \in \mathcal{U}, b \models q\}$ is an open subset of $S_y(M)$.
- μ is *Borel-definable* over M if it is M-invariant and the above set is a Borel set of $S_y(M)$.

If μ is finitely satisfiable in M then it is M-invariant. Also μ is definable over M if and only if it is M-invariant and for any open (resp. closed) subset X of $[0,1]$, the set $\{q \in S_y(M) : \mu(\phi(x;b)) \in X \text{ for any } b \in \mathcal{U}, b \models q\}$ is an open (resp. closed) subset of $S_y(M)$.

A measure μ is M-invariant (resp. finitely satisfiable in M), if and only if all types in $S(\mu)$ are M-invariant (resp. finitely satisfiable in M). This uses the fact that if μ is M-invariant, then $\mu(\phi(x;b)\triangle\phi(x;b')) = 0$ whenever $b \equiv_M b'$, which follows from Proposition 7.15. On the other hand, definability of measures is a new phenomenon that does not translate to types.

LEMMA 7.17. *Let $\mu \in \mathfrak{M}_x(\mathcal{U})$ and $M \prec \mathcal{U}$.*

(i) *If μ is smooth over M, then μ is finitely satisfiable and definable in M.*
(ii) *If μ is M-invariant and smooth, then it is smooth over M.*

PROOF. (i): Let $\phi(x;b)$ be such that $\mu(\phi(x;b)) > 0$. Then by Lemma 7.8, taking ε small enough, there is a formula $\theta^0(x) \in L(M)$ such that $\models \theta^0(x) \to \phi(x;b)$ and $\mu(\theta^0(x)) > 0$. In particular, $\theta^0(x)$ is consistent, and therefore has a point in M. It follows that also $\phi(x;b)$ has a point in M.

Let $r \in [0,1]$ be such that $\mu(\phi(x;b)) < r$. Take $\varepsilon > 0$ so that $\mu(\phi(x;b)) < r - 2\varepsilon$. Lemma 7.8 applied to $\phi(x;y)$ and ε gives us formulas $\psi_i^\varepsilon(y) \in L(M)$, $i < n$ which partition of y-space. One of them, say $\psi_i^\varepsilon(y)$, is satisfied by b. Then for any b', $\psi_i^\varepsilon(b')$ implies that $\mu(\phi(x;b')) \leq r - \varepsilon < r$. This proves definability of μ.

(ii): Assume that μ is M-invariant and smooth. Let $N \succ M$ be $|M|^+$ saturated such that μ is smooth over N. Fix a formula $\phi(x;y)$ and $r > 0$. Then the set $\{q \in S_y(N) : \mu(\phi(x;b)) \leq r \text{ for any } b \models q\}$ is an closed set of $S_y(N)$. Its projection to $S_y(M)$ is also closed, hence μ is definable over M.

Take now some $\varepsilon > 0$. Lemma 7.8 applied to $\phi(x;y)$ and ε gives us formulas $\psi_i(y;d)$, $\theta_i^0(x;d)$ and $\theta_i^1(x;d)$, $i < n$, where we have made the parameters explicit. By definability of μ, there is an M-formula $\zeta(z)$ satisfied by d and such that for any $d' \in \zeta(\mathcal{U})$, the formulas $\psi_i(y;d')$, $\theta_i^0(x;d')$ and $\theta_i^1(x;d')$, $i < n$ also satisfy the conclusion of the lemma. Then we can find such a d' in M. It follows that μ is smooth over M. ⊣

LEMMA 7.18. *Let $p \in S(\mathcal{U})$ be some M-invariant type, then p is Borel-definable over M.*

PROOF. Let $\phi(x;y)$ be a formula. We need to see that the set $\{b \in \mathcal{U} : p \vdash \phi(x;b)\}$ is Borel over M. Let $b \in \mathcal{U}$. Take a maximal N such that there is a sequence $(a_i : i < N) \models p^{(N)}|_M$ with

$$\neg(\phi(a_i;b) \leftrightarrow \phi(a_{i+1};b)) \text{ for } i < N-1. \qquad (*)$$

By the discussion on eventual types in Section 2.2.3, the formula $\phi(a_{N-1};b)$ holds if and only if $p \vdash \phi(x;b)$.

Let $A_N(b)$ say that there are $(a_i : i < N)$ such that $(*)$ holds and $\models \phi(a_{N-1};b)$. Similarly define $B_N(b)$ to hold if there is $(a_i : i < N)$ such that $(*)$ holds and $\models \neg\phi(a_{N-1};b)$. Then both $A_N(x)$ and $B_N(x)$ are type-definable sets. We have $p \vdash \phi(x;b)$ if and only if there is some integer $N \leq \text{alt}(\phi)+1$ such that $A_N(x) \wedge \neg B_{N+1}(x)$ holds.

This shows that $\{b \in \mathcal{U} : p \vdash \phi(x;b)\}$ is a finite Boolean combination of type-definable (over M) sets. ⊣

PROPOSITION 7.19. *Let $\mu \in \mathfrak{M}(\mathcal{U})$ be M-invariant, then μ is Borel-definable over M.*

PROOF. This follows easily from the previous lemma and Proposition 7.11 which says that a measure can be approximated by types in its support. ⊣

We now have all we need to define the product of an invariant measure over another, similarly as we did for types. So let $\mu(x) \in \mathfrak{M}(\mathcal{U})$ be M-invariant and let $\lambda(y) \in \mathfrak{M}(\mathcal{U})$ be any measure. We define $\omega(x, y) = \mu(x) \otimes \lambda(y)$ as a global measure in two variables by the formula:

$$\omega_{xy}(\phi(x, y; b)) = \int_{q \in S_y(N)} f(q) d\lambda_y|_N,$$

where N is a small model containing Mb and $f : q \mapsto \mu(\phi(x, d; b))$ for some (any) $d \models q$. Note that f is a measurable function by Borel-definability of μ. We need to see that this definition does not depend on the choice of N. So let $N_1 \subseteq N_2$ be two models containing Mb and let f_1, f_2 be the associated functions. Let $\pi : S(N_2) \to S(N_1)$ be the canonical restriction map. Then $f_2 = \pi^*(f_1)$. For any clopen set $B \subseteq N_1$, $\lambda|_{N_1}(B) = \lambda|_{N_2}(\pi^{-1}(B))$, hence this holds also when B is Borel and then by approximating f_1 by step functions, we deduce that the integrals of f_1 and f_2 are the same.

Note that if $\mu = p$ and $\lambda = q$ are types, then we recover the usual product $p(x) \otimes q(y)$.

If λ is itself M-invariant, then $\mu(x) \otimes \lambda(y)$ is M-invariant (since the whole construction is invariant under automorphisms fixing M).

Exercise 7.20. If both μ and λ are finitely satisfiable in M (resp. M-definable), then $\mu(x) \otimes \lambda(y)$ is finitely satisfiable in M (resp. M-definable).

We now argue that the product of measures is associative. Let $\mu(x), \eta(y)$ and $\lambda(z)$ be three global measures, and assume that μ and η are both M-invariant. Let $\phi(x, y, z) \in L(M)$, let $v = ((\mu_x \otimes \eta_y) \otimes \lambda_z)(\phi(x, y, z))$ and $w = (\mu_x \otimes (\eta_y \otimes \lambda_z))(\phi(x, y, z))$. We want to show that $v = w$.

First assume that $\mu = p$ is a type. Let $d\phi(y, z)$ denote the Borel subset of $S_{yz}(M)$ defined by $r \in d\phi \iff p \vdash \phi(x, b, c)$ where $(b, c) \models r$. Then $w = \eta_y \otimes \lambda_z(d\phi(y, z)) = \int_{q \in S_z(M)} f(q) d\lambda_z$, where $f : q \to \eta(d\phi(y, c))$, for some $c \models q$. But this is exactly v.

Now if μ is an arbitrary invariant measure, then given a formula $\phi(x, y, z)$ and ε, we can find a measure $\tilde{\mu}$, which is an average of types such that $|\mu(\phi(x, b, c)) - \tilde{\mu}(\phi(x, b, c))| \leq \varepsilon$ for all b, c. Let $\tilde{v}, \tilde{w}$ be the corresponding v and w with μ replaced by $\tilde{\mu}$. Then $\tilde{v}$ and $\tilde{w}$ are respectively at distance ε from v and w. By the previous paragraph, $\tilde{v} = \tilde{w}$, and we conclude that $v = w$.

If μ is an M-invariant global measure, then we can define by induction $\mu^{(1)}(x) = \mu(x)$ and $\mu^{(n+1)}(x_0, \dots, x_n) = \mu(x_0) \otimes \mu^{(n)}(x_1, \dots, x_n)$. And of course $\mu^{(\omega)}(x_0, x_1, \dots)$ is the union of $\mu^{(n)}(x_0, \dots, x_{n-1})$ for $n < \omega$.

This construction is different from the product measure in probability theory in that the space of n-types is not the n-fold product of the space of 1-types. Hence $\mu^{(n)}$ is not defined on the same space and carries more information than a usual product measure would. However, $\mu^{(n)}$ and the measure-theoretic

n-fold product of μ agree on sets which are measurable for the product algebra. In other words, they agree on formulas which are Boolean combinations of formulas in one variable. This will allow us to use probability theoretic methods (mainly the law of large numbers) with $\mu^{(n)}$ playing the role of the product measure.

We now prove that a finitely satisfiable measure and a definable one commute (the case of types was done in Lemma 2.23).

LEMMA 7.21. *Let $\mu(x), \lambda(y)$ be two global M-invariant measures. Assume that $\mu(x) \otimes p(y) = p(y) \otimes \mu(x)$ for any $p \in S_y(\mathcal{U})$ in the support of λ. Then μ and λ commute.*

PROOF. Let $\phi(x, y)$ be a formula, and write as in the definition $(\mu_x \otimes \lambda_y)(\phi(x, y)) = \int_{q \in S_y(M)} f(q) d\lambda_x$. Fix $\varepsilon > 0$. Approximate that integral up to ε by a finite sum $\sum_{i=1,\dots,n} \lambda(X_i)c_i$, where $X_i = \{q \in S_y(M) : \mu(\phi(x, b)) \in [r_i, t_i]$, for some $b \models q\}$ with $r_i, t_i \in [0, 1]$. By Proposition 7.11, we can find types $q_1, \dots, q_m$ weakly random for λ, such that if $\tilde{\lambda}$ denotes the average $\frac{1}{m}\sum q_j$, then:

- $\tilde{\lambda}(X_i)$ is within ε of $\lambda(X_i)$, for all $i \leq n$;
- $\tilde{\lambda}(\phi(a, y))$ is within ε of $\lambda(\phi(a, y))$ for all $a \in \mathcal{U}$.

The first condition ensures that $\mu_x \otimes \tilde{\lambda}_y(\phi(x, y))$ is within ε of $\mu_x \otimes \lambda_y(\phi(x, y))$ and the second one ensures that $\tilde{\lambda}_y \otimes \mu_x(\phi(x, y))$ is within ε of $\lambda_y \otimes \mu_x(\phi(x, y))$. As μ commutes with $\tilde{\lambda}$, the result follows. ⊣

PROPOSITION 7.22. *Let $\mu(x) \in \mathfrak{M}(\mathcal{U})$ be definable and $\lambda(y) \in \mathfrak{M}(\mathcal{U})$ be finitely satisfiable, then $\mu(x) \otimes \lambda(y) = \lambda(y) \otimes \mu(x)$.*

PROOF. By the previous lemma, we may assume that $\lambda(y) = q(y)$ is a type. Let M such that both μ and q are M-invariant. Assume for a contradiction that there is $r \in [0, 1]$ and $\varepsilon > 0$ such that $\mu_x \otimes q_y(\phi(x; y)) < r - 2\varepsilon$ whereas $q_y \otimes \mu_x(\phi(x; y)) > r + 2\varepsilon$. By definability of μ, there is a formula $\psi(y) \in L(M)$ such that $q \vdash \psi(y)$ and for all $b \in \psi(\mathcal{U})$, we have $\mu(\phi(x; b)) < r - \varepsilon$. Also, by Borel-definability of q, there is a Borel set $X \subseteq S_x(M)$ such that $q \vdash \phi(a; y)$ if and only if $\text{tp}(a/M) \in X$.

Pick $p_1, \dots, p_n \in S_x(M)$ such that for all $b \in M$, $\text{Av}(p_1, \dots, p_n; \phi(x; b))$ is within ε of $\mu(\phi(x; b))$ and also $\text{Av}(p_1, \dots, p_n; X)$ is within ε of $\mu(X)$. Realize each p_i by $a_i \in \mathcal{U}$. By finite satisfiability of q, find some $b_0 \in \psi(M)$ such that $\models \phi(a_i; b_0) \iff q \vdash \phi(a_i; y)$ for all i.

By choice of p_i's and $\psi(y)$, we have $\text{Av}(p_1, \dots, p_n; \phi(x; b_0)) < r$. On the other hand, by the choice of b_0, that quantity is equal to $\frac{1}{n}\sum_i q(\phi(a_i; y)) = \frac{1}{n}|\{i : a_i \in X\}|$ which is within ε of $\mu(X) = q_y \otimes \mu_x(\phi(x; y)) > r + 2\varepsilon$. Contradiction. ⊣

7.5. Generically stable measures

Similarly as we did for types, we define a *generically stable measure* as being a global measure which is both definable and finitely satisfiable (in some small model M). We will prove an analog of Theorem 2.29. First we define a new notion.

DEFINITION 7.23. Let $\mu(x)$ be a global M-invariant measure. We say that μ is *fim* (frequency interpretation measure) if for any formula $\phi(x;y) \in L$ and $\varepsilon > 0$, there is a family $(\theta_n(x_1, \dots, x_n) : n < \omega)$ of formulas in $L(M)$ such that:

- $\lim \mu^{(n)}(\theta_n(x_1, \dots, x_n)) = 1$;
- for n big enough, for any $(a_1, \dots, a_n) \in \theta_n(\mathcal{U})$, and any $b \in \mathcal{U}$, $\mathrm{Av}(a_1, \dots, a_n; \phi(x;b))$ is within ε of $\mu(\phi(x;b))$.

We will see later that a measure is fim if and only if it is generically stable. Left to right is easy, but the converse requires more work.

The main result needed is an adaptation of the VC-theorem from measure theory to the context of Keisler measures.

VC-type results. We call a measure $\mu(x_1, \dots, x_n)$ over A *symmetric* if for any permutation σ of $\{1, \dots, n\}$ and any formula $\phi(x_1, \dots, x_n) \in L(A)$, we have $\mu(\phi(x_1, \dots, x_n)) = \mu(\phi(x_{\sigma 1}, \dots, x_{\sigma n}))$. This definition extends naturally to measures in infinitely many variables.

Let $\phi(x;y)$ be a formula. We fix some integer n and write $\bar{x} = (x_1, \dots, x_n)$ and $\bar{x}' = (x'_1, \dots, x'_n)$. We define

$$f'_n(\bar{x}, \bar{x}'; b) = |\,\mathrm{Av}(x_1, \dots, x_n; \phi(x;b)) - \mathrm{Av}(x'_1, \dots, x'_n; \phi(x;b))|$$

and

$$f_n(\bar{x}, \bar{x}') = \sup_{b \in \mathcal{U}} f'_n(\bar{x}, \bar{x}'; b).$$

Note that for any $\varepsilon > 0$ the statement $f'_n(\bar{x}, \bar{x}; b) > \varepsilon$ can be expressed as a first order formula $\theta_{n,\varepsilon}(\bar{x}, \bar{x}'; b)$. Similarly, the statement $f_n(\bar{x}, \bar{x}') > \varepsilon$ can be expressed as $(\exists b)\theta_{n,\varepsilon}(\bar{x}, \bar{x}'; b)$.

LEMMA 7.24. *Fix a formula $\phi(x;y)$ and $\varepsilon > 0$. Let $\mu_n(x_1, \dots, x_n, x'_1, \dots, x'_n)$ be a symmetric measure over $\emptyset$. Then with notations as above we have*

$$\mu_n(f_n(\bar{x}, \bar{x}') > \varepsilon) \le 4\pi_\phi(n) \exp\left(-\frac{n\varepsilon^2}{8}\right).$$

PROOF. This is the analogue of Claim 2 inside the proof Theorem 6.6. The proof is exactly the same: we just have to make the necessary translations and check that everything still makes sense.

Let $G = \{-1, 1\}^n$. If $\phi(x;y)$ is a formula, we let $\ulcorner\phi(a;b)\urcorner$ be equal to 1 it the sentence $\phi(a;b)$ holds and 0 otherwise.

CLAIM. *We have*

$$\mu_n(f_n(\bar{x},\bar{x}')>\varepsilon)\leq\frac{1}{2^{n-1}}\sum_{\sigma\in G}\mu_n\left(\left\{\bar{x}:\sup_b\frac{1}{n}\left|\sum_{i=1}^{n}\sigma_i\ulcorner\phi(x_i;b)\urcorner\right|>\varepsilon/2\right\}\right).$$

PROOF OF CLAIM. Call D_σ the set of which we take the μ_n-measure on the right-hand side. Note that it is $\emptyset$-definable. For $\sigma\in G$ let $\theta'_\sigma(\bar{x},\bar{x}';y)$ be the formula expressing that

$$\left|\frac{1}{n}\sum_{i=1}^{n}\sigma_i\left(\ulcorner\phi(x_i;y)\urcorner-\ulcorner\phi(x'_i;y)\urcorner\right)\right|>\varepsilon$$

and let $\theta_\sigma(\bar{x},\bar{x}')=(\exists b)\theta'_\sigma(\bar{x},\bar{x}';b)$. By symmetry of μ_n, for any $\sigma\in G$ we have $\mu_n(\theta_\sigma(\bar{x},\bar{x}'))=\mu_n(f_n(\bar{x},\bar{x}')>\varepsilon)$. Hence

$$\mu_n\left(f_n(\bar{x},\bar{x}')>\varepsilon\right)=\frac{1}{2^n}\sum_{\sigma\in G}\mu_n\left(\theta_\sigma(\bar{x},\bar{x}')\right).$$

The claim then follows exactly as Claim 1 in Theorem 6.6. ⊣ (Claim)

The translation of the rest of the proof poses no difficulty either. For a fixed tuple $\bar{a}$ of length n and $b\in\mathcal{U}$, we let $A_b(\bar{a})$ be the set of $\sigma\in G$ for which we have $\frac{1}{n}\left|\sum_i\sigma_i\ulcorner\phi(a_i;b)\urcorner\right|>\varepsilon/2$. Then Chernoff's bound gives

$$\frac{1}{2^n}|A_b(\bar{a})|\leq 2\exp\left(-\frac{n\varepsilon^2}{8}\right).$$

The tuple $\bar{a}$ being fixed, there are only $\pi_\phi(n)$ possible values for the set $A_b(\bar{a})$ as b varies. Hence the cardinality of the set $A_*=\bigcup_b A_b(\bar{a})$ is at most $e=2^{n+1}\pi_\phi(n)\exp\left(-n\varepsilon^2/8\right)$. This means that a tuple $\bar{a}$ can be in at most e many sets D_σ; thus we can bound the sum $\sum_{\sigma\in G}\mu_n(D_\sigma)$ by e and the lemma follows. ⊣

COROLLARY 7.25. *Let $\mu(x_1,\dots)$ be a Keisler measure in ω variables over some set A. Assume that $\mu|_\emptyset$ is symmetric. Let $\phi(x;y)\in L$ and fix $\varepsilon>0$. Then there is n such that $\mu(\exists y(f'_n(x_1,\dots,x_n,x_{n+1},\dots,x_{2n};y)>\varepsilon))\leq\varepsilon$.*

PROOF. Apply Lemma 7.24 letting $\mu_n(x_1,\dots,x_{2n})$ be the restriction of μ to the variables $(x_1,\dots,x_{2n})$. ⊣

PROPOSITION 7.26. *Let μ be a global M-invariant measure and assume that $\mu^{(\omega)}(x_1,\dots)|_\emptyset$ is symmetric. Then μ is fim.*

PROOF. Fix a formula $\phi(x;y)$ and $\varepsilon>0$ small enough. Write $\bar{a}=(a_1,\dots,a_n)$ and $\bar{a}'=(a'_1,\dots,a'_n)$. Define $\theta_n(\bar{a},\bar{a}')$ to say that for all b, $\mathrm{Av}(a_1,\dots,a_n;\phi(x;b))$ is within $\varepsilon/4$ of $\mathrm{Av}(a'_1,\dots,a'_n;\phi(x;b))$. By the previous corollary, for n big enough we have $\mu^{(2n)}(\theta_n(x_1,\dots,x_{2n}))>1-\varepsilon$. In particular, there is $\bar{a}\in\mathcal{U}$, such that we have $\mu^{(n)}(\theta_n(\bar{a},x_1,\dots,x_n))>1-\varepsilon$.

Now fix $b \in \mathcal{U}$ and let $\zeta_n(x_1, \ldots, x_n)$ say that $\text{Av}(x_1, \ldots, x_n; \phi(x; b))$ is within $\varepsilon/4$ of $\mu(\phi(x; b))$. The law of large numbers (B.4) gives

$$\mu^{(n)}(\zeta_n(x_1, \ldots, x_n)) \geq 1 - 4/n\varepsilon^2.$$

Taking n large enough, we have $\mu^{(n)}(\zeta_n(x_1, \ldots, x_n)) \geq 1/2$. It follows that $\mu^{(n)}(\theta_n(\bar{a}, x_1, \ldots, x_n) \wedge \zeta_n(x_1, \ldots, x_n)) > 0$. In particular the value $\text{Av}(a_1, \ldots, a_n; \phi(x; b))$ is within $\varepsilon/2$ of $\mu(\phi(x; b))$.

To conclude observe that for any $\bar{a}'$ satisfying $\theta_n(\bar{a}, x_1, \ldots, x_n)$ and any b, $\text{Av}(a'_1, \ldots, a'_n; \phi(x; b))$ is within ε of $\mu(\phi(x; b))$. As $\mu^{(n)}(\theta_n(\bar{a}, x_1, \ldots, x_n)) > 1 - \varepsilon$, we have what we want. ⊣

We can do the same incorporating a Borel set X.

PROPOSITION 7.27. *Let μ be a global M-invariant measure, and assume that $\mu^{(\omega)}(x_1, \ldots)|_\emptyset$ is symmetric. Then for any Borel set X, formula $\phi(x; y)$ and $\varepsilon > 0$, there are $a_1, \ldots, a_n \in \mathcal{U}$ such that for all $b \in \mathcal{U}$, $\text{Av}(a_1, \ldots, a_n; \phi(x; b) \cap X)$ is within ε of $\mu(\phi(x; b) \cap X)$.*

PROOF. We need to go through the proof of Lemma 7.24 and of Proposition 7.26 replacing everywhere $\phi(x; b)$ by $\phi(x; b) \cap X$. Sets which were previously definable are now Borel, but this does not create any difficulty. ⊣

Remark 7.28. The statements above imply that if μ is A-invariant and $\mu^{(\omega)}$ is symmetric, then μ is equal to the average of any *realization* of $\mu^{(\omega)}|_A$. More precisely, let $\lambda(x_1, \ldots)$ be a global measure which extends $\mu^{(\omega)}|_A$. Let $\phi(x; b) \in L(\mathcal{U})$ be a formula and let $r = \mu(\phi(x; b))$. Fix $\varepsilon > 0$, then for all but finitely many values of n, we have $\lambda(\phi(x_n; b)) \in (r - \varepsilon, r + \varepsilon)$.

This implies in particular that μ is determined by $\mu^{(\omega)}|_A$. (This is also true for arbitrary invariant measures, but we do not have the tools to prove it.)

PROOF. Assume that we have $\lambda(\phi(x_n; b)) - r > \varepsilon$ for infinitely many values of n. Without loss, this is true for all n. Now set $\omega = \mu^{(\omega)}(x'_1, x'_2, \ldots) \otimes \lambda(x_1, x_2, \ldots)$. Then $\omega|_\emptyset$ is symmetric. By Corollary 7.25, for n large enough, we have $\omega(f'_n(x_1, \ldots, x_n, x'_1, \ldots, x'_n; b) < \varepsilon/4) > 1/2$.

Let $\theta_n(x_1, \ldots, x_n)$ say $\text{Av}(x_1, \ldots, x_n; \phi(x; b)) - r > \varepsilon/2$. Then by the weak law of large numbers, we have $\lambda(\theta_n(x_1, \ldots, x_n)) \to 1$. Letting the formula $\theta'_n(x'_1, \ldots, x'_n)$ say $|\text{Av}(x'_1, \ldots, x'_n; \phi(x; b)) - r| < \varepsilon/4$, then we also have $\mu^{(\omega)}(\theta'_n(x'_1, \ldots, x'_n)) \to 1$. It follows that $\omega(\theta_n(x_1, \ldots, x_n) \wedge \theta'_n(x'_1, \ldots, x'_n)) \to 1$. But this contradicts what we established in the previous paragraph. ⊣

Properties of generically stable measures.

THEOREM 7.29. *Let μ be a global M-invariant measure. Then the following are equivalent:*

(i) *μ is generically stable;*
(ii) *for any formula $\phi(x; y) \in L$ and $\varepsilon > 0$, there are $a_1, \ldots, a_n \in M$ such that for any $b \in M$, $\text{Av}(a_1, \ldots, a_n; \phi(x; b))$ is within ε of $\mu(\phi(x; b))$.*

(iii) *μ is fim;*
(iv) *$\mu^{(\omega)}(x_1,\dots)|_M$ is symmetric;*
(v) *μ commutes with itself: $\mu(x) \otimes \mu(y) = \mu(y) \otimes \mu(x)$.*

PROOF. (i) $\Rightarrow$ (v): Follows from Proposition 7.22.
(v) $\Rightarrow$ (iv): Clear by associativity of $\otimes$.
(iv) $\Rightarrow$ (iii) is Proposition 7.26.
(iii) $\Rightarrow$ (ii): Clear.
(ii) $\Rightarrow$ (i): Easy. ⊣

PROPOSITION 7.30. *Let $\mu(x)$ be a generically stable measure, then for any invariant measure $\lambda(y)$, we have $\mu(x) \otimes \lambda(y) = \lambda(y) \otimes \mu(x)$.*

PROOF. By Lemma 7.21, we may assume that $\lambda_y = q_y$ is an invariant type. Let M be such that both μ and q are invariant over M and let $N \supset M$ be $|M|^+$-saturated. Take some $b \models q|_N$ in $\mathcal{U}$ and let $\phi(x,y) \in L(M)$ be a formula. Let $X \subseteq S_x(M)$ be the set of types p such that $q \vdash \phi(a,y)$ for some (any) $a \models p$. Then, by definition, we have $\mu_x \otimes q_y(\phi(x,y)) = \mu(\phi(x,b))$ and $q_y \otimes \mu_x(\phi(x,y)) = \mu(X)$.

Fix $\varepsilon > 0$. By Proposition 7.27 there are $a_1,\dots,a_n \in N$ such that for all $b' \in \mathcal{U}$, $\mathrm{Av}(a_1,\dots,a_n;\phi(x;b'))$ is within ε of $\mu(\phi(x;b'))$ and also $\mathrm{Av}(a_1,\dots,a_n;X)$ is within ε of $\mu(X)$. Then $\mu_x \otimes q_y(\phi(x,y))$ is equal to $\mathrm{Av}(a_1,\dots,a_n;X)$ hence within ε of $\mathrm{Av}(a_1,\dots,a_n;\phi(x;b))$, which by definition of X is within ε of $q_y \otimes \mu_x(\phi(x,y))$.

As this holds for all $\varepsilon > 0$, the result follows. ⊣

PROPOSITION 7.31. *Let $\mu(x)$ be generically stable and A-invariant. Then μ is the unique A-invariant extension of $\mu|_A$.*

PROOF. Assume that $\lambda(x)$ is a global A-invariant extension of μ. As in the proof of Proposition 2.35, we show that $\mu^{(\omega)}|_A = \lambda^{(\omega)}|_A$. It follows that λ is also generically stable, and by the remark after Proposition 7.27, $\mu = \lambda$. ⊣

Example 7.32. Recall the examples of measures given in Example 7.2. We go through them again, and see that most of them are generically stable.

- If $p \in S_x(\mathcal{U})$ is a global type, then p is generically stable as a measure if and only if it is generically stable as a type. More generally, an average $\sum_{i<\omega} a_i p_i$ of types is generically stable if and only if all the p_i's are generically stable (assuming of course $a_i > 0$ for all i).
- Any Borel probability measure on $\mathbb{R}$ induces a smooth Keisler measure. See [63]. In fact, more generally, under mild assumptions, if M is a model equipped with a σ-algebra making externally definable sets measurable and μ_0 is a σ-additive measure on M, then μ_0 induces a generically stable Keisler measure. See [113, Theorem 5.1] for a precise statement and a proof.
- The average measure of an indiscernible sequence $I = (a_i : i \in [0,1])$ is always generically stable. It is easy for example to check fim. It is

smooth if and only if the sequence I satisfies a property called *distality*. See Proposition 9.30.
- Let $M = \prod_{\mathcal{D}} M_n$ be an ultraproduct of finite structures. Define μ_n as the normalized counting measure on M_n and let μ be an ultralimit of the μ_n's. Then μ is generically stable. Indeed the VC-theorem implies that property (ii) in Theorem 7.29 holds for μ_n where the number of points in the approximation depends only on $\phi(x; y)$ and ε. Hence this is also true at the limit.

Exercise 7.33. Let μ be the average of the indiscernible sequence $I = (a_i : i <\in [0, 1])$. Given $n < \omega$ and a formula $\phi(x_1, \dots, x_n) \in L(\mathcal{U})$ give an explicit formula for $\mu^{(n)}(\phi(x_1, \dots, x_n))$ in terms of the sequence (a_i). Check that $\mu^{(n)}$ is symmetric, which gives another proof that μ is generically stable.

Exercise 7.34. Let $\mu(x)$ be M-invariant and generically stable. Let $\phi(x) \in L(\mathcal{U})$ such that $\mu(\phi(x)) > 0$. Then there is some $\psi(x) \in L(M)$ such that $\psi(M) \subseteq \phi(M)$ and $\mu(\psi(x)) > 0$.

References and related subjects

Keisler measures were introduced in the context of NIP theories by Keisler in [71]. There, he defines a notion of *smooth measure* which is weaker than the one we consider here. In [60], Hrushovski, Peterzil and Pillay revive those ideas and use measures to study groups definable in o-minimal theories; in particular, the concept of an *invariant measure* of a group is central in their approach. The boundedness properties come from that work. Those ideas are developed further in Hrushovski and Pillay [61], where in particular Borel definability is observed. Finally, generically stable measures are defined and studied in Hrushovski, Pillay and Simon [63].

Ben Yaacov proves in [17] that the *randomization* of an NIP theory is NIP. This statement is equivalent to the following: Let $\omega(x_1, \dots)$ be a measure such that for any formula $\theta(x_1, \dots, x_n)$, and tuples $i_1 < \cdots < i_n$, $j_1 < \cdots < j_n$, $\omega(\theta(x_{i_1}, \dots, x_{i_n})) = \omega(\theta(x_{j_1}, \dots, x_{j_n}))$. Then for any formula $\phi(x; b)$ and ε, there can be only finitely many indices n for which $|\omega(\phi(x_n; b)) - \omega(\phi(x_{n+1}; b))| \geq \varepsilon$. This result is thus the analog of Lemma 2.7 for measures and could be used to translate in a uniform way proofs from types to measures. We have done without it here in order to keep this text self-contained.

Further results about *generically stable measures* can be found in Hrushovski, Pillay and Simon [62] and Simon [113].

Chapter 8

DEFINABLY AMENABLE GROUPS

An important class of examples of Keisler measures are translation-invariant measures on definable groups. They can serve as a substitute for generic types used in the stable setting. However, unlike in the stable situation, not all NIP groups admit an invariant measure. Groups that do are called *definably amenable*. The class of definably amenable groups is very diverse, as the following three examples illustrate: The free group $\mathbb{F}_2$ is definably amenable because it is stable. Any solvable group is definably amenable, because any such group is amenable as a discrete group. The group $SO_3(\mathbb{R})$ is definably amenable because it admits a normalized Haar measure for which all definable sets are measurable.

As we will see, properties of the invariant measure—generic stability and smoothness—translate into properties of the group.

NIP is assumed throughout this chapter.

8.1. Connected components

Our main interest is in definable groups. However, everything in this section generalizes without difficulties to type-definable groups, so we will work in that context.

A type-definable group is a type-definable set $G = \bigcap_i \phi_i(x)$ equipped with a definable map $\cdot_G$ such that $(G, \cdot_G)$ is a group (when interpreted in the monster model, or equivalently in any sufficiently saturated model containing the required parameters). When no confusion can arise, we will drop the index G in the notation $\cdot_G$.

By compactness, we see that:

(i) there is some formula $\phi(x)$ containing G such that for any a, b, c satisfying $\phi(x)$, we have $(a \cdot b) \cdot c = a \cdot (b \cdot c)$ and $a \cdot 1_G = 1_G \cdot a = a$;

(ii) for every formula $\phi_0(x)$ containing G, there is some formula $\phi_1(x)$ containing G such that for any a, b in $\phi_1(x)$, $a \cdot b \models \phi_0(x)$ and for all a in $\phi_0(x)$ there is a unique b in $\phi_1(x)$ which is an inverse for a.

Iterating this last point to obtain $\phi_2(x), \phi_3(x), \ldots$, we see that the intersection $\bigcap_{k<\omega} \phi_k(x)$ is a type-definable group containing G. This shows that G is the intersection of type-definable groups each defined by a countable intersection of formulas.

Let M be a model over which G is defined. The space of types $p(x)$ over M which concentrate on G (that is such that $p \vdash x \in G$) will be denoted by $S_G(M)$. Similarly, $\mathcal{M}_G(M)$ is the space of measures $\mu(x)$ which concentrate on G, i.e., such that $\mu(G) = 1$.

The group $G(M)$ acts on the left on $\mathcal{M}_G(M)$ by $g \cdot \mu(\phi(x)) = \mu(\phi(g \cdot x))$. This action restricts to an action on $S_G(M)$: for $p \in S_G(M)$ and $a \models p$, $g \cdot p = \operatorname{tp}(g \cdot a/M)$. A measure $\mu \in \mathcal{M}_G(M)$ is said to be (left) G-invariant if $g \cdot \mu = \mu$ for all $g \in G(M)$.

There is also an action on the right of $G(M)$ on $\mathcal{M}_G(M)$ defined by $(\mu \cdot g)(\phi(x)) = \mu(\phi(x \cdot g))$.

Example 8.1. A typical example of a type-definable group is the subgroup of $(\mathbb{Z}, +)$ of elements divisible by every positive integer n. In the standard model, this group reduces to the identity, but in a saturated model it is infinite.

Another example is the group of infinitesimal elements in (a saturated model of) $(\mathbb{R}; +, <)$ defined as the intersection of the intervals $[-1/k, 1/k]$ for $k \in \mathbb{N}^*$.

Assumption: Throughout this section, G is some $\emptyset$-type-definable group (we will repeat this assumption in the statement of some theorems). Recall that NIP is always assumed.

The reader is welcome to think of G as a definable group. She will not lose much by doing so.

8.1.1. Bounded index subgroups. By a *relatively definable* subset of G, we mean the intersection of G with a definable set X. Let H be a relatively definable subgroup of G and we are interested in the index $[G : H]$.

Consider first the case where G is definable. If the index $[G(M) : H(M)]$ is finite for some model M, then this index does not depend on the model M we take (as long as H is defined over it). We say that H has finite index in G. If this is not the case, then by compactness, the index $[G(M) : H(M)]$ can be as big as we want, and we say that H has infinite index in G.

When G is type-definable, there is a caveat: it may happen that the index $[G(M) : H(M)]$ is abnormally small for some model M (in fact, it may well be that over M, $G(M)$ reduces to the identity). Hence one must take M to be sufficiently saturated. Then the same dichotomy holds: either for some sufficiently saturated M the index $[G(M) : H(M)]$ is finite, in which case its value does not depend on the choice of M, or this index is infinite in which case it may be made as large as we want by increasing M.

Let us now consider the case of a type-definable subgroup H. There are again two cases: either the index $[G(M) : H(M)]$ can be made as large as we want by increasing M, in which case we say that H has unbounded index in G, or for some model M, the quotient $G/H(M)$ does not increase as we increase M. We then say that H has bounded index in G.

In fact the same dichotomy is true if H is merely invariant over some set A (meaning that for any $a, b \in \mathcal{U}$, if $a \equiv_A b$, then $a \in H \iff b \in H$). In this case, the equivalence relation $a \equiv_H b$ defined on G by $ab^{-1} \in H$ is A-invariant. The subgroup H has bounded index in G if and only if the relation $\equiv_H$ is a bounded equivalence relation in the sense of Section 5.1 (when restricted to G). In particular, H has unbounded index if and only if there is an indiscernible sequence $(a_i : i < \omega)$ of elements of G which lie in pairwise distinct H-cosets.

Exercise 8.2. Show that if H is a type-definable subgroup of G of finite index, then it is relatively definable.

8.1.2. G^0. A family $(H_i)_{i \in A}$ of subgroups of G is uniformly (relatively) definable if for some formula $\phi(x; y)$, for every i, there is b_i such that $H_i = G \cap \phi(x; b_i)$.

We recall the Baldwin-Saxl theorem 2.13. The statement was made in the case of a definable group G, but the proof works just as well if G is type-definable.

THEOREM 8.3 (Baldwin-Saxl). *Let G be a type-definable group and let $(H_i)_{i \in A}$ be a uniformly (relatively) definable family of subgroups of G. Then there is an integer N such that for any finite intersection $\bigcap_{i \in A} H_i$, there is a subset $A_0 \subseteq A$ of size N with $\bigcap_{i \in A} H_i = \bigcap_{i \in A_0} H_i$.*

So for any formula $\phi(x; y) \in L$, x a variable of the same sort as G, there is an integer N_ϕ such that any finite intersection of subgroups of the form $G \cap \phi(x; b)$ is equal to a subintersection of size N. Given k, this implies that any finite intersection of subgroups of the form $G \cap \phi(x; b)$ and of index $\leq k$ in G has index $\leq kN$. But then this is also true for an infinite intersection. Therefore there is a definable subgroup $G^0_{\phi,k}$ of index at most kN which is the intersection of all subgroups of index $\leq k$ definable by an instance of $\phi(x; y)$. The subgroup $G^0_{\phi,k}$ is relatively definable and invariant under all automorphisms. Therefore it is relatively definable over $\emptyset$. (One could also give an explicit definition over $\emptyset$.)

Let $G^0_\phi = \bigcap_{k<\omega} G^0_{\phi,k}$. It is a type-definable (over $\emptyset$) subgroup of G of bounded index. Finally, we let $G^0 = \bigcap_\phi G^0_\phi$. It is again a type-definable over $\emptyset$ subgroup of bounded index and is equal to the intersection of all relatively definable subgroups of finite index. As the class of definable groups of finite index is stable under conjugation, G^0 is a normal subgroup of G.

In stable theories, G^0 is called the connected component of G. The terminology originates from algebraic geometry, where G^0 denotes the smallest algebraic subgroup of G of finite index. For groups definable in NIP theories, G^0 plays a less central role as there are two other connected components which come into play: G^{00} and G^∞ which we define below.

8.1.3. G^{00}. Let A be a small set of parameters. We denote by G^{00}_A the intersection of all type-definable over A subgroups of bounded index. It is itself a type-definable over A subgroup of bounded index and hence is the smallest such. It may happen that G^{00}_A does not depend on A, and hence is equal to $G^{00}_\emptyset$. In this case, we call it G^{00} and say that G^{00} exists. Hence G^{00}, when it exists, is the smallest type-definable (over any set of parameters) subgroup of bounded index in G. Like G^0, it is a normal subgroup of G.

It turns out that G^{00} always exists in NIP theories.

THEOREM 8.4. *Let G be an $\emptyset$-type-definable group, then G^{00} exists.*

PROOF. Assume that G^{00} does not exist. Then we can find an arbitrarily large collection $(H_i : i < \kappa)$ of pairwise distinct type-definable subgroups of bounded index. We may assume that each H_i is defined as an intersection of at most $\aleph_0$ formulas. By Ramsey and compactness, there is such a sequence $(H_i : i < \omega)$ which is indiscernible, by which we mean that there is a type-definable set $\Phi(x; \bar{y})$ and an indiscernible sequence $(\bar{b}_i : i < \omega)$ such that $H_i = \Phi(x; \bar{b}_i)$. Without loss, for any $\bar{b}$, $\Phi(x; \bar{b})$ is a type-definable subgroup.

CLAIM. *H_i does not contain the intersection $\bigcap_{j \neq i} H_j$.*

PROOF OF CLAIM. Assume it did and insert in place of H_i a very long sequence $(H'_l : l < \kappa')$ such that the whole sequence is still indiscernible. Then each H'_l contains the intersection $\bigcap_{j \neq i} H_j$. But that intersection has bounded index, hence there are only boundedly many subgroups of G containing it. Taking κ' large enough, we have a contradiction. ⊣ (Claim)

For each $i < \omega$, let $a_i \in \cap_{j \neq i} H_j \setminus H_i$ chosen so that the sequence $(a_i, \bar{b}_i : i < \omega)$ is indiscernible. Then there is a formula $\phi(x; \bar{y})$ implied by $\Phi(x; \bar{y})$ such that $\phi(a_i; \bar{b}_j)$ holds if and only if $i \neq j$. Let $\theta(x; \bar{y})$ be a formula in $\Phi(x; \bar{y})$ such that $\models \bigwedge_{i<3} \theta(x_i; \bar{y}) \to \phi(x_0 \cdot x_1 \cdot x_2; \bar{y})$, where $\cdot$ is the group operation. For $I = \{i_1, \ldots, i_n\} \subset \omega$ a finite subset, define $a_I = a_{i_1} \cdot a_{i_2} \cdot \ldots \cdot a_{i_n}$. Then we have $\theta(a_I; \bar{b}_i)$ if and only if $i \notin I$. (If $i \notin I$, this is clear, otherwise observe that any a_{i_k} can be written as $c_0 \cdot a_I \cdot c_1$ with $c_0, c_1 \in H_{i_k}$.) This shows that the formula $\theta(x; \bar{y})$ has IP. ⊣

Example 8.5. Let R be a real closed field and consider the usual circle group $S_1 \subset R^2$ of elements (x, y) such that $x^2 + y^2 = 1$. Then $S_1^0 = S_1$ and S_1^{00} is the set of infinitesimal of S_1, that is the set of points $(x, y) \in S_1$ infinitely close to $(1, 1)$.

The quotient G/G^{00} is in canonical bijection with the standard group $S_1(\mathbb{R})$. We will see later that it is in fact homeomorphic to it, when endowed with an adequate topology.

8.1.4. G^∞. Finally, we consider invariant subgroups of bounded index.

Given a set A of parameters, we define G^∞_A as the smallest A-invariant subgroup of G of bounded index. If this group does not depend on A, then we call it G^∞ and say that G^∞ exists. Hence G^∞, when it exists, is a $\emptyset$-invariant subgroup of bounded index which for any A contains all A-invariant subgroups of bounded index. It is also a normal subgroup of G.

As observed earlier, if H is any A-invariant subgroup then the relation $x \equiv_H y$ defined by $xy^{-1} \in H$ is an A-invariant equivalence relation. If H has bounded index, then $\equiv_H$ is a bounded equivalence relation. This implies that if a and b have the same Lascar strong type over A, then $a \equiv_H b$ and hence $ab^{-1} \in H$. Hence H contains the subgroup generated by $\{a \cdot b^{-1} : \mathrm{Lstp}(a/A) = \mathrm{Lstp}(b/A)\}$. Conversely that latter subgroup is A-invariant and has bounded index. Hence it is equal to G^∞_A. In particular, if M is a model, then G^∞_M is precisely the subgroup generated by

$$X_{\equiv_M} := \{a \cdot b^{-1} : \mathrm{tp}(a/M) = \mathrm{tp}(b/M)\}.$$

If $\Phi(x;\bar{y})$ is a set of formulas, where $\bar{y}$ has the size of M, we will write $X_{\equiv^\Phi_M} = \{a \cdot b^{-1} : \phi(a;M) = \phi(b;M)$, for any $\phi \in \Phi\}$, where we assume that we have fixed an enumeration of M.

We also use the standard notation $X^n = \{x_1 \cdot \ldots \cdot x_n : x_1, \ldots, x_n \in X\}$.

LEMMA 8.6. *Let A be any set of parameters and $c \in G(\mathcal{U})$, then we have* $c\,(X_{\equiv_A})\,c^{-1} \subseteq (X_{\equiv_A})^2$.

PROOF. Let $a \equiv_A b$ and consider $cab^{-1}c^{-1}$. There is $d \in G(\mathcal{U})$ such that $\mathrm{tp}(a,c/A) = \mathrm{tp}(b,d/A)$. Then $cab^{-1}c^{-1} = (ca)(db)^{-1} \cdot dc^{-1} \in (X_{\equiv_A})^2$. ⊣

THEOREM 8.7. *Let G be a $\emptyset$-type-definable group, then G^∞ exists.*

PROOF. Assume not. Then for some small model M, $G^\infty_M \neq G^\infty_\emptyset$. Let $\lambda = |M|$. The intersection $\bigcap_{M'} G^\infty_{M'}$, where M' ranges over all models of size λ is a $\emptyset$-invariant subgroup of G. If it has bounded index, then G^∞_M contains $G^\infty_\emptyset$, hence is equal to it, which contradicts the hypothesis. It follows that we can find an arbitrary long sequence $(M_i : i < \kappa)$, each M_i being of size λ such that $G^\infty_{M_i}$ does not contain the intersection $\bigcap_{j<i} G^\infty_{M_j}$. By Erdős-Rado, we can find such a sequence $(M_i : i < \omega)$ which is moreover indiscernible (where we have implicitly fixed an enumeration of each M_i). For each $i < \omega$, let $c_i \in \bigcap_{j \neq i} G^\infty_{M_j} \setminus G^\infty_{M_i}$ and assume that $(M_i, c_i : i < \omega)$ is indiscernible. In particular, for some $m < \omega$, for all $j < i$, $c_i \in \left(X_{\equiv_{M_j}}\right)^m$. By compactness, find some finite set $\Phi(x;\bar{y})$ of formulas such that

$$c_i \notin (X_{\equiv^\Phi_{M_i}})^{m+4}. \tag{1}$$

By the previous lemma, for some finite set $\Phi'(x;\bar{y})$, for all $c \in G(\mathcal{U})$, we have $c\left(X_{\equiv^{\Phi'}_{M_i}}\right)c^{-1} \subseteq \left(X_{\equiv^\Phi_{M_i}}\right)^2$.

Now to any finite sequence $I = (i_1, \dots, i_n)$ of distinct elements of ω, we associate

$$c_{I,0} = c_{2i_1+1} \cdot \dots \cdot c_{2i_n+1};$$
$$c_{I,1} = c_{2i_1} \cdot \dots \cdot c_{2i_n}.$$

Let $j < \omega$.

CLAIM 1. *If $j \notin I$, then $c_{I,0} \cdot c_{I,1}^{-1} \in X_{\equiv_{M_{2j}}} \subseteq X_{\equiv^{\Phi'}_{M_{2j}}}$.*

PROOF OF CLAIM 1. If $j \notin I$, then by indiscernibility of our sequence, $c_{I,0} \equiv_{M_{2j}} c_{I,1}$ and hence $c_{I,0} \cdot c_{I,1}^{-1} \in X_{\equiv_{M_{2j}}}$. ⊣ (Claim 1)

CLAIM 2. *If $j \in I$, then $c_{I,0} \cdot c_{I,1}^{-1} \notin X_{\equiv^{\Phi'}_{M_{2j}}}$.*

PROOF OF CLAIM 2. Assume that $j \in I$ and the conclusion does not hold. Write $I = I_1 + (j) + I_2$. We then have:

$$\begin{aligned} c_{I,0} \cdot c_{I,1}^{-1} &= c_{I_1,0} \cdot c_{2j+1} \cdot c_{I_2,0} \cdot c_{I_2,1}^{-1} \cdot c_{2j}^{-1} \cdot c_{I_1,1}^{-1} \\ c_{I,0} \cdot c_{I,1}^{-1} \cdot c_{I_1,1} \cdot c_{2j} &= c_{I_1,0} \cdot c_{2j+1} \cdot c_{I_2,0} \cdot c_{I_2,1}^{-1} \\ c_{2j} &= c_{I_1,1}^{-1} \cdot c_{I,1} \cdot c_{I,0}^{-1} \cdot c_{I_1,0} \cdot c_{2j+1} \cdot c_{I_2,0} \cdot c_{I_2,1}^{-1} \\ &= [c_{I_1,1}^{-1}(c_{I,1}c_{I,0}^{-1})c_{I_1,1}] \cdot (c_{I_1,1}^{-1}c_{I_1,0}) \cdot c_{2j+1} \cdot (c_{I_2,0}c_{I_2,1}^{-1}). \end{aligned}$$

By assumption, $c_{2j+1} \in \left(X_{\equiv_{M_{2j}}}\right)^m$. We also know that $c_{I_1,1}^{-1}(c_{I,1}c_{I,0}^{-1})c_{I_1,1} \in c_{I_1,1}^{-1} X_{\equiv^{\Phi'}_{M_{2j}}} c_{I_1,1} \subseteq \left(X_{\equiv_{M^{\Phi}_{2j}}}\right)^2$ and both $c_{I_1,1}^{-1}c_{I_1,0}$ and $c_{I_2,0}c_{I_2,1}^{-1}$ are in $X_{\equiv_{M_{2j}}}$. So in total, we have $c_{2j} \in \left(X_{\equiv^{\Phi}_{M_{2j}}}\right)^{m+4}$ which contradicts (1). ⊣ (Claim 2)

By claims 1 and 2, the formula

$$\Psi(x; \bar{y}) = \exists x_1 x_2 \left(\bigcap_{\phi \in \Phi'} (\phi(x_1; \bar{y}) \leftrightarrow \phi(x_2; \bar{y})) \wedge x = x_1 x_2^{-1} \right)$$

has IP. ⊣

Exercise 8.8. Show that in an arbitrary theory, if G^∞ exists then so does G^{00}.

To summarize, given a $\emptyset$-type-definable group G, we have defined three connected components: $G \supseteq G^0 \supseteq G^{00} \supseteq G^\infty$. In general, all of these inclusions can be strict. We have given examples when $G \neq G^0$ and $G^{00} \neq G^0$. It is more difficult to give an example where $G^\infty \neq G^{00}$. For that, we refer to Conversano and Pillay [30].

8.1.5. Compact quotients. Let E be a type-definable over $\emptyset$ equivalence relation on $\mathcal{U}$ (for simplicity on singletons). Recall from Section 5.1 that E is said to be bounded if $\mathcal{U}/E$ has small cardinality, equivalently if for some (every) small model M, $a \equiv_M b$ implies aEb. Assume from now on that this holds. Let $\pi : \mathcal{U} \to \mathcal{U}/E$ be the quotient map. We endow $\mathcal{U}/E$ with the *logic topology* defined as follows: a set $F \subseteq \mathcal{U}/E$ is closed if and only if $\pi^{-1}(F)$ is type-definable over some (any) model M.

LEMMA 8.9. *The space $\mathcal{U}/E$ equipped with the logic topology is a compact Hausdorff space.*

PROOF. Let M be any small model, then we have a map $f : S(M) \to \mathcal{U}/E$ which assigns to each type p the unique E-class on which it concentrates. By definition of the topology on $\mathcal{U}/E$, this map is continuous. Hence $\mathcal{U}/E$ is the image of a compact space, and as such it is compact.

We are left with showing that the space is Hausdorff. Let $a, b \in \mathcal{U}$ such that $\neg aEb$. Then for every two points x and y, we have $xEa \wedge yEb \Longrightarrow \neg xEy$. By compactness, there is a formula $\phi(x, y)$ such that $\models xEy \to \phi(x, y)$ and $\mathcal{U} \models \phi(x, a) \wedge \phi(y, b) \to \neg xEy$. Let $O_a = \{x \in \mathcal{U}/E : \pi^{-1}(x) \subseteq \phi(\mathcal{U}, a)\}$ and $O_b = \{y \in \mathcal{U}/E : \pi^{-1}(y) \subseteq \phi(\mathcal{U}, b)\}$. Then O_a and O_b are disjoint open neighborhoods of $\pi(a)$ and $\pi(b)$ respectively. ⊣

If G is a definable, or type-definable, group then we have on G the equivalence relation $xG^{00} = yG^{00}$ of being in the same G^{00} coset. It is type-definable and bounded. As above, we equip the quotient space G/G^{00} with the logic topology, making it a compact Hausdorff space. In fact the topology is even compatible with the group structure.

LEMMA 8.10. *The group G/G^{00} equipped with the logic topology is a compact topological group.*

PROOF. One has to check that the group operation and the inverse map are continuous. We leave the verification as an exercise to the reader. ⊣

Hence to any type-definable group G, we have associated a canonical compact group G/G^{00}.

Example 8.11.
- If $G^{00} = G^0$ then G/G^{00} is a profinite group: it is the inverse image of the groups G/H, where H ranges over all relatively definable subgroups of finite index.
- If $G = (\mathbb{Z}, +)$, then $G^{00} = G^0$ is the set of elements divisible by all n. The quotient G/G^{00} is isomorphic as a topological group to $\hat{\mathbb{Z}} = \varprojlim \mathbb{Z}/n\mathbb{Z}$.
- If $G = S_1$ is the circle group defined in a real closed field R, then G^{00} is the set of infinitesimal elements of G and G/G^{00} is canonically isomorphic to the standard circle group $S_1(\mathbb{R})$.
- More generally, if $G = G(R)$ is a semialgebraic group defined over $\mathbb{R}$ such that $G(\mathbb{R})$ is a compact Lie group, then again G/G^{00} is isomorphic to $G(\mathbb{R})$. (Even more generally, if G is any *definably compact* group

defined in an o-minimal structure, then G/G^{00} is a Lie group: this is part of the content of Pillay's conjecture—now a theorem. See Peterzil [88] for a survey.)

- This does not hold any more if G is a non-compact Lie group. For example if $G = (\mathbb{R}, +)$, then $G^{00} = G$ and G/G^{00} is trivial.

8.2. Definably amenable groups

From now on, we assume that G is a $\emptyset$-definable group (and not merely type-definable as before).

Recall that a discrete group G_0 is *amenable* if there is a finitely additive probability measure defined on all subsets of G_0. It is well known that any solvable group is amenable.

DEFINITION 8.12. The definable group G is *definably amenable* if there is a global Keisler measure $\mu(x)$ concentrating on G which is left invariant, that is such that $\mu(\phi(g \cdot x)) = \mu(\phi(x))$ for all $g \in G(\mathcal{U})$.

A few important observations:

- If G is definably amenable, then it also admits a global measure which is right invariant. In fact, if $\mu(x)$ is left invariant, then $\nu(x)$ defined by $\nu(\phi(x)) = \mu(\phi(x^{-1}))$ is right invariant. However it is not true in general that a left invariant measure is also right invariant (Exercise 8.24).
- If for some model M there is a left invariant Keisler measure μ_0 on M-definable sets, then G is definably amenable. This can be seen for example by taking a saturated elementary extension of $\tilde{M}_{\mu_0}$ as defined after Lemma 7.4 and letting μ be the measure obtained from it.
- If G admits a left invariant type, that is a global type p such that $g \cdot p = p$ for all $g \in G$, then it is definably amenable. This is the case for example for $G = (\mathbb{R}; +)$ which has two invariant types at $\pm\infty$. If for some model M, there is a left invariant type p_0 over M, then there is one over any model and in particular over $\mathcal{U}$ (to go down, take a restriction, and to go up, take an heir).

Example 8.13 (Amenable groups). If for some model M, the group $G(M)$ is amenable as a discrete group, then G is definably amenable. In particular, any abelian group and more generally any solvable group is definably amenable.

Example 8.14 (Stable groups). Let G be a stable definable group. A definable set $X \subseteq G$ is *left generic* if finitely many left translates of X cover G. Similarly for right generic. One shows that left generic sets coincide with right generic sets and that there always are generic types, that is types that concentrate only on generic sets. In fact the natural map $S_G(\mathcal{U}) \to G/G^0$ induces a homeomorphism between G/G^0 and the set T of global generic types. In particular, any generic type is invariant by translation by elements of G^0.

It follows that G is definably amenable: If G is connected, that is $G^0 = G$, then it admits a unique generic type, which is left and right invariant. In general, the Haar measure on G/G^0 translates to a left invariant Keisler measure concentrating on T.

For example, the free group $\mathbb{F}_2$, which is known to be stable by the work of Sela, is definably amenable. However $\mathbb{F}_2$ is not amenable as an abstract group.

Example 8.15 (Compact Lie groups). Let $G(\mathbb{R})$ be a *compact* semialgebraic Lie group. Then, seen as a definable group in RCF, G is definably amenable: the compact group $G(\mathbb{R})$ admits a normalized Haar measure h, and all definable subsets of G are measurable with respect to h.

In particular, the group $SO_3(\mathbb{R})$ is definably amenable, although it is known not to be amenable as a discrete group (essentially the content of the Banach-Tarski paradox).

This does not work if G is not compact since then the Haar measure has infinite weight. In fact we show in Example 8.22 that the definable group $G = SL_n(\mathbb{R})$ is not definably amenable. A characterization of definably amenable groups in o-minimal expansions of RCF is given in Conversano and Pillay [30].

PROPOSITION 8.16. *If G is definably amenable and A is an extension base over which G is defined, then there is a global measure μ which is left-translation invariant and does not fork over A.*

PROOF. Assume not. Let Inv_G be the subset of $\mathfrak{M}_G(\mathcal{U})$ consisting of left invariant measures. It is a non-empty closed subspace, and therefore compact. If it does not contain a measure which is non-forking over A, then by compactness, there are formulas $\psi_0(x;b), \ldots, \psi_{n-1}(x;b) \in L(\mathcal{U})$, each forking over A, and $\varepsilon > 0$ such that for every $\mu \in Inv_G(\mathcal{U})$, for some i, $\mu(\psi_i(x;b)) > \varepsilon$. In particular, if $\psi(x;b)$ denotes the disjunction of the ψ_i's, then for every $\mu \in Inv_G$, $\mu(\psi(x;b)) > \varepsilon$.

As A is an extension base, the formula $\psi(x;b)$ divides over A. Let $(b_i : i < \omega)$, $b_0 = b$, be A-indiscernible such that $\{\psi(x;b_i) : i < \omega\}$ is inconsistent. As $Inv_G \subset \mathfrak{M}_G(\mathcal{U})$ is setwise invariant under automorphisms of $\mathcal{U}$ fixing A, every $\mu \in Inv_G$ satisfies $\mu(\psi(x;b_i)) > 0$ for each $i < \omega$. By Lemma 7.5 the partial type $\{\psi(x;b_i) : i < \omega\}$ is consistent, contradicting dividing. ⊣

DEFINITION 8.17. A global type $p \in S_G(\mathcal{U})$ is (left) f-generic over A if no left translate of p forks over A.

LEMMA 8.18. *Assume that G admits a global f-generic type p (over some set A). Then $G^\infty = G^{00}$ and p is G^{00}-invariant (by translation on the left).*

PROOF. Let M be such that p is f-generic over M. Let $Stab_l(p)$ be the left stabilizer of p, that is the set of $g \in G$ such that $g \cdot p = p$. It is an M-invariant subgroup of G. We will show that $Stab_l(p) = \{g_1 \cdot g_2^{-1} : g_1 \equiv_M g_2\}$.

Assume that $g_1, g_2 \in G$ have the same type over M and we show that $g_1 \cdot p = g_2 \cdot p$. Let $\phi(x; b) \in L(\mathcal{U})$ and $f \in \mathrm{Aut}(\mathcal{U}/M)$ mapping g_1 to g_2. Then $g_1 \cdot p \vdash \phi(x; b) \iff f(g_1) \cdot p \vdash \phi(x; f(b)) \iff g_2 \cdot p \vdash \phi(x; b)$. The last equivalence uses the fact that $g_2 \cdot p$ is M-invariant. It follows that $g_1 \cdot g_2^{-1} \in Stab_l(p)$.

Conversely, let $h \in Stab_l(p)$ and let $a \models p|_{Mh}$. Then $h \cdot a \equiv_M a$ and $h = (h \cdot a) \cdot a^{-1}$ as required.

We conclude that $Stab_l(p) = \{g_1 \cdot g_2^{-1} : g_1 \equiv_M g_2\}$. In particular the latter set is a subgroup. We know that it generates G^∞, hence it is equal to it. Since it is also type-definable, we conclude that $Stab_l(p) = G^\infty = G^{00}$. ⊣

We now present a very useful construction used to turn a definable group G into an automorphism group. So let G be $\emptyset$-definable and fix some model M. We add to the structure M a new sort S. The universe of S is a copy of $G(M)$. We do not put any structure on S except for a relation $R(x; y, z)$ which holds for $x \in G$ and $y, z \in S$ if and only if $z^{-1} \cdot y = x$. Fixing some $b \in S$, the function associating to any $x \in G$ the unique $y \in S$ such that $\models R(x; y, b)$ is then a definable bijection between G and S. However there is no $\emptyset$-definable bijection. In fact, for any $b, b' \in S$, there is a (unique) automorphism of the full structure fixing M pointwise and sending b to b', namely $y \mapsto b'b^{-1} \cdot y$.

Note that this construction is conservative, that is does not add any new definable sets in M. Also, if $b \in S$ is a point and $A \subset M$, then any subset X of M definable over Ab is definable over A (X is also definable over Ab' for any $b' \in S$, by the same formula and then we can quantify on b' to obtain a formula over A).

LEMMA 8.19. *Let G be a definable group and construct S as above. Then G admits a global f-generic type over A if and only if the formula $x_S = x_S$, x_S of sort S, does not fork over A.*

PROOF. Assume that $x_S = x_S$ does not fork over A and let $p_S(x_S)$ be a global non-forking extension. Fix an arbitrary point $b \in S$. It induces a definable bijection between S and G which sends $p_S(x_S)$ to some global type $p(x)$. The type p does not fork over Ab. By the remarks above, it follows that p does not fork over A. Changing b to b' amounts to translating p by $b^{-1}b'$, therefore no translate of p forks over A.

Conversely, assume that p is f-generic over A. Any point b in S gives rise to a bijection between G and S. If q is a type in G, we let q_b be the type in the sort S which is the image of q under this bijection. Take now two points b, b' in S. Then p_b does not fork over Ab. However, p_b is also equal to $q_{b'}$ where q is the adequate translate of p. Hence it does not fork over Ab'. As b' was arbitrary, p_b does not fork over A. ⊣

COROLLARY 8.20. *If G admits an f-generic type over some model M, then it admits an f-generic type over any model M_0.*

PROOF. We work in the expanded structure with a new sort S. Assume that the formula $x_S = x_S$ does not fork over M, but does fork over M_0. By definition of forking, $x_S = x_S$ implies a disjunction $\bigvee_{i<n} \phi_i(x_S; b^i)$, each $\phi_i(x_S; b^i)$ divides over M_0. Let $(b^i_j : j < \omega)$ be an M_0-indiscernible sequence witnessing dividing. Let M_1 realize a coheir of $\text{tp}(M/M_0)$ over $M_0 \cup \{b^i_j : i < n, j < \omega\}$. Then each sequence $(b^i_j : j < \omega)$ remains indiscernible over M_1 and $\phi_i(x_S; b_i)$ divides over M_1. But since M_1 and M have the same type over $\emptyset$, this implies that $x_S = x_S$ forks over M. Contradiction. ⊣

PROPOSITION 8.21. *Let M be a model over which G is defined. Then G admits a global f-generic type over M if and only if G is definably amenable.*

PROOF. We may restrict to countable L and then by Corollary 8.20, we may assume that M is also countable.

One direction follows from Proposition 8.16: if G is definably amenable then it admits a global left invariant measure μ which does not fork over M and any type weakly random for μ is f-generic over M.

For the converse, let $p(x)$ be a global f-generic type. Then by Lemma 8.18, p is G^{00}-invariant. In particular, if $\bar{g} \in G/G^{00}$, then the translate $\bar{g} \cdot p$ is well defined. Let h be the Haar measure on G/G^{00}. We define a global measure $\mu(x)$ by $\mu(\phi(x)) = h(\{\bar{g} \in G/G^{00} : \bar{g} \cdot p \vdash \phi(x)\})$, for $\phi(x) \in L(\mathcal{U})$. We need to check that this is allowed, that is that the set on the right is Borel. Assume this for now, then G-invariance of μ follows from the invariance of the Haar measure. Also if $\mu(\phi(x)) > 0$, then $\phi(x)$ is in some translate of p, hence does not fork over M. Therefore any type weakly random for μ is f-generic over M.

It remains to check that μ is well defined. Let $\phi(x) \in L(\mathcal{U})$. By Borel definability of p (Proposition 7.19), the set $X = \{g \in G : g \cdot p \vdash \phi(x)\}$ is Borel over M. In fact we know from Proposition 7.19 that it is a finite Boolean combination of closed sets. In particular, as L and M are countable, it is a countable union of closed sets. The canonical map $\pi : S_G(M) \to G/G^{00}$ is continuous, hence closed. We conclude that $\pi(X)$ is Borel in G/G^{00} as required. ⊣

Example 8.22 ($SL_n(\mathbb{R})$). We show that the group $SL_n(\mathbb{R})$ is not definably amenable. First note that since $PSL_n(K) = SL_n(K)/\{\pm 1\}$ is simple for any infinite field K, we know that there are no normal subgroups of bounded index (definable or not) except possibly a normal subgroup of index 2. The fact that $SL_n(\mathbb{R})$ is perfect rules out this latter case, hence $G^{00} = G^0 = G$. It follows that any f-generic type needs to be G-invariant. Let $p(x)$ be such a type. Then p must imply either $|x_{11}| \geq |x_{21}|$ or $|x_{11}| < |x_{21}|$. Assume the first case. Let $g \in SL_n(\mathbb{R})$ be the matrix with 1's on the diagonal and t, $|t| > 2$, on the

$(2, 1)$-entry. Then $g \cdot p \vdash |x_{11}| < |x_{21}|$ contradicting invariance. In the other case, do the same but with t in the $(1, 2)$-entry.

Exercise 8.23. Let G be definable and H a definable normal subgroup of G.

1. If G is definably amenable, then so is G/H.
2. If both H and G/H are definably amenable, then so is G.

Exercise 8.24.
1. Let $G = \mathbb{R} \rtimes \mathbb{Z}/2\mathbb{Z}$ where the non-zero element of $\mathbb{Z}/2\mathbb{Z}$ acts on $\mathbb{R}$ by $x \mapsto -x$. Show that, as a definable group in RCF, G has a left invariant measure which is not right invariant.
2. Show that any definably amenable group admits a global Keisler measure which is both left and right invariant.

Exercise 8.25.
1. Let $G = (\mathbb{Q}^2, +)$ as a definable set in $(\mathbb{Q}; 0, <)$. Show that G has $2^{\aleph_0}$ global left invariant types.
2. Let $G = (\mathbb{R}^2, +)$ seen as a definable set in RCF. Then G has unboundedly many left invariant types. In particular it admits left invariant types which are not invariant over any small model.

Exercise 8.26 (Automorphism groups). Let $p \in S(A)$ be a type. Show that p does not fork over A if and only if p extends to a global A-invariant measure.

[*Hint*: Assume that p has a global non-forking extension $\tilde{p}$. Then $\tilde{p}$ is invariant under $G = \operatorname{Aut}(bdd(A)/A)$, which is a compact group. Using the Haar measure of G we can average translates of $\tilde{p}$ and generate an A-invariant Keisler measure as in Proposition 8.21. See [61, Proposition 4.7] for details.]

8.3. fsg groups

DEFINITION 8.27. Let G be a $\emptyset$-definable group. We say that G has *fsg* (finitely satisfiable generics) if there is a small model M_0 and a global type p such that every left translate of p is finitely satisfiable in M_0.

In particular, such a type p is f-generic over M_0 and hence G is definably amenable.

DEFINITION 8.28. A definable set $X \subseteq G$ is *left generic* if finitely many left translates of X cover G. We define right generic similarly.

PROPOSITION 8.29. *Let G be a definable fsg group. Let p, M_0 be as in the definition of fsg. For X a definable subset of G, the following are equivalent*:

(i) *X is left generic*;
(ii) *X is right generic*;
(iii) *every left (right) translate of X meets $G(M_0)$*;
(iv) *some left (right) translate of X is in p.*

PROOF. Assume that X is left generic. Then X is in some left translate of p hence X meets $G(M_0)$. The same is true starting with any left translate of X instead of X. By compactness, there are finitely many points $g_1, \dots, g_n \in G(M_0)$ such that any left translate of X meets one the g_i's. This exactly means that $X \cdot g_1^{-1} \cup \dots \cup X \cdot g_n^{-1}$ covers G, hence X is right generic. ⊣

Let G have fsg, witnessed by p. By the previous proposition, left and right generic sets of G coincide, so we just call them generics. The type p is generic in the sense that all definable sets in it are generic.

Let μ_p be the left invariant measure on G obtained from averaging translates of p as in Proposition 8.21. Then μ_p is generic in the sense that only generic sets have positive measure. Conversely, if X is generic then it must have positive measure with respect to any invariant measure. Hence X is generic if and only if $\mu_p(X) > 0$. This shows that non-generic formulas form an ideal: if $X_1 \cup X_2$ is generic, then either X_1 or X_2 is generic. Also any generic formula extends to a global generic type.

LEMMA 8.30. *Let G be a definable fsg group. A generic definable subset X of G has a point in $G(M)$ for any small model M.*

PROOF. Let $\phi(x; a) \in L(\mathcal{U})$ be a formula such that $X = \phi(G; a)$ is generic in G: some m translates of X cover G. We know that for any a' such that m translates of $\phi(G; a')$ cover G, $\phi(x; a')$ intersects $G(M_0)$. Hence by compactness there is $\{g_1, \dots, g_n\} \subset G(M_0)$ which intersects any such $\phi(x; a')$. This is a definable condition in $(g_1, \dots, g_n)$, hence we can find such points in any model. ⊣

LEMMA 8.31. *Let G be a $\emptyset$-definable group and M_0 a small model. Let $\lambda_0 \in \mathfrak{M}_G(M_0)$ be a measure over M_0 invariant by left translation by elements of $G(M_0)$. Then λ_0 admits a global extension λ which is left invariant and definable.*

PROOF. First we find some model $M_1 \succ M_0$ and an extension λ_1 of λ_0 to M_1 which is left invariant (by translation by elements of $G(M_1)$) and such that λ_1 admits a unique left invariant extension to any bigger model. This is done by a smoothing argument as in Proposition 7.9: Notice first that any left invariant measure over some M admits a left invariant extension to any larger model (for example using the $\tilde{M}_\mu$-construction exposed after Lemma 7.4). Then if there are two such extensions, we consider their average and iterate. This process must stop as in the construction of smooth measures.

Having obtained λ_1, let λ be the unique global left invariant extension of λ_1. We claim that λ is definable. Using again the construction referred to in the previous paragraph, let $\tilde{M}_1 = (M_1, [0,1], f_\phi)_\phi$ encode the measure λ_1. Let N be an elementary extension of M_1 and $\phi(x; a) \in L(N)$. Let $r = \lambda(\phi(x; a))$ and fix some $\varepsilon > 0$. By assumption, there is no elementary extension of $\tilde{M}_1$ containing a whose associated measure ν satisfies $|\nu(\phi(x; a)) - r| \geq \varepsilon$. By compactness, there is some formula $\psi(y) \in \text{tp}(a/M)$ such that any $a' \models \psi(y)$ has the same property. In particular, any $a' \in \psi(\mathcal{U})$ satisfies $|\lambda(\phi(x; a')) - r| < \varepsilon$ which proves definability of λ. ⊣

PROPOSITION 8.32. *Let G be a definable fsg group, then G admits a unique global left invariant measure μ. Moreover μ is generic, is the unique right invariant measure and is generically stable.*

PROOF. Start by letting $\mu = \mu_p$ be the global generic left invariant measure obtained from p as above. Then μ is finitely satisfiable is any small model. Let λ be any global right invariant measure on G which is definable over some M. By Proposition 7.22, μ and λ commute. Let $\phi(x) \in L(M)$ be a formula and we compute $\mu_x \otimes \lambda_y(\phi(y \cdot x))$.

On the one hand, we have (with some abuse of notations): $\mu_x \otimes \lambda_y(\phi(y \cdot x)) = \int_{b \in \mathcal{U}} \mu(\phi(b \cdot x)) d\lambda(b) = \int_b \mu(\phi(x)) d\lambda(b) = \mu(\phi(x))$. On the other hand, using commutativity of μ and λ: $\mu_x \otimes \lambda_y(\phi(y \cdot x)) = \int_{a \in \mathcal{U}} \lambda(\phi(y \cdot a)) d\mu(a) = \int_a \lambda(\phi(y)) d\mu(a) = \lambda(\phi(y))$.

Hence $\mu = \lambda$. This shows that μ is right invariant and definable, hence it is generically stable. It is in fact the unique definable right invariant measure. As μ is right invariant, we can do the same reasoning starting with some left invariant λ, and we conclude that μ is also the unique definable left invariant measure on G.

Now let λ be any global left (or right) invariant measure. Let M_0 be a small model. By the previous lemma, $\lambda|_{M_0}$ extends to a global definable invariant measure, which must be equal to μ. Therefore $\lambda = \mu$. ⊣

PROPOSITION 8.33. *Let G be a definable group. Then G has fsg if and only if it is definably amenable and admits a generically stable left invariant measure.*

PROOF. One direction follows from the previous statement. For the converse, assume that G is definably amenable and admits a generically stable left invariant measure μ. Then in particular μ is finitely satisfiable in some model M. Let p be a weakly random type for μ. By invariance of μ, all left translates of p are also weakly random for μ. Hence p and all its translates are finitely satisfiable in M which shows that G has fsg. ⊣

Example 8.34.
- Every stable group has fsg. (A generic type of a stable group does not fork over $\emptyset$ and by stability it is finitely satisfiable in any small model.)
- Any compact semialgebraic Lie group $G(\mathbb{R})$ has fsg.

  [*Sketch of proof*: As G is compact, there is a standard part map $\mathrm{st} : G \to G(\mathbb{R})$. Let h denote the normalized Haar measure on $G(\mathbb{R})$. For X a definable subset of G, $\mathrm{st}(X)$ is Borel and we can define $\mu(X) = h(\mathrm{st}(X))$. If the dimension of X is less than that of G, then $\mu(X) = 0$. Hence if $\mu(X) > 0$, also $\mu(\mathring{X}) > 0$ and then $\mathring{X}$ contains a point in $G(\mathbb{R})$. Thus μ is finitely satisfiable in the model $\mathbb{R}$ and G has fsg. This is an instance of *compact domination* which is studied in the next section.]

More generally, any definably compact group definable in an o-minimal expansion of a real closed field has fsg (see Hrushovski, Peterzil and Pillay [60]).
- The additive group $(\mathbb{Z}_p, +)$ of p-adic integers has fsg, as one can check directly.

8.4. Compact domination

Let X be a type-definable over M set. Let K be a compact set equipped with a Borel probability measure h. We say that a map $f : X \to K$ is definable over M if for every closed subset $C \subseteq K$, the preimage $f^{-1}(C)$ is type-definable over M. In other words, f induces a continuous map from $S_X(M)$ (the set of types concentrating on X) to K.

We say that X is *compactly dominated* by (K, h, f) if for every $\mathcal{U}$-definable set D, the set $\{a \in K : f^{-1}(a) \cap D \neq \emptyset \text{ and } f^{-1}(a) \cap D^c \neq \emptyset\}$ has h-measure 0.

When this holds, the measure h lifts to an M-invariant global Keisler measure μ on X defined by $\mu(D) = h(f(D))$ for any $\mathcal{U}$-relatively definable set D. In particular, h is the pushforward of μ by f.

LEMMA 8.35. *Let X be type-definable and μ be an M-invariant Keisler measure concentrating on X. Let μ_0 denote the associated Borel measure on $S(M)$. Then X is compactly dominated by $(S(M), \mu_0, \text{tp}(\cdot/M))$ if and only if μ is smooth.*

PROOF. Assume first that μ is smooth. Let $b \in \mathcal{U}$ and fix a formula $\phi(x; b) \in L(Mb)$. By smoothness of μ, for any $\varepsilon > 0$ we can find sets $\theta_{0,\varepsilon}(x)$ and $\theta_{1,\varepsilon}(x)$ both over M such that $\mathcal{U} \models \theta_{0,\varepsilon}(x) \to \phi(x; b) \to \theta_{1,\varepsilon}(x)$ and $\mu(\theta_{1,\varepsilon}(x)) - \mu(\theta_{0,\varepsilon}(x)) < \varepsilon$. Let $C_\phi(x) = \bigwedge_\varepsilon \theta_{1,\varepsilon}(x) \wedge \neg\theta_{0,\varepsilon}(x)$, where ε ranges over $(0, 1) \cap \mathbb{Q}$. Then $C_\phi(x)$ is closed over M and has μ-measure 0. For $a \in X(\mathcal{U}) \setminus C_\phi(\mathcal{U})$, for any $a' \equiv_M a$, we have $\models \phi(a; b) \leftrightarrow \phi(a'; b)$. This proves domination.

Conversely, assume that X is dominated by $(S(M), \mu_0, \text{tp}(\cdot/M))$. Let $\phi(x; b) \in L(\mathcal{U})$ be a formula. Let $D_1 \subseteq S(M)$ be the set of types p such that $p(x) \vdash \phi(x; b)$ and $D_2 \subseteq S(M)$ the set of types p such that $p(x) \vdash \neg\phi(x; b)$. Then any extension $\tilde{\mu}$ of μ must satisfy $\mu(D_1) \leq \tilde{\mu}(\phi(x; b)) \leq 1 - \mu(D_2)$. By hypothesis $\mu(S(M) \setminus (D_1 \cup D_2)) = 0$. Hence $\mu(D_1) = 1 - \mu(D_2)$ and $\tilde{\mu}$ is determined. ⊣

Let now G be a definable group. We say that G is compactly dominated (as a group) by some (K, π) if K is a compact group, $\pi : G \to K$ a definable morphism and G is compactly dominated (as a type-definable set) by (K, h, π) where h is the Haar measure on K.

LEMMA 8.36. *If G is compactly dominated by some (K, π), then it is compactly dominated by $(G/G^{00}, \pi_0)$ where π_0 is the canonical projection.*

PROOF. Let $H = \pi^{-1}(e)$. It is a type-definable subgroup of G and we have a bijection between G/H and K. Hence H has bounded index in G. Therefore G^{00} is contained in H and π factors through π_0. The result follows. ⊣

With this lemma in mind, we will say that a definable group G is *compactly dominated*, if it is compactly dominated by $(G/G^{00}, \pi_0)$.

The aim of this section is to prove the following theorem.

THEOREM 8.37. *The definable group G is compactly dominated if and only if it admits a smooth left invariant measure.*

By Proposition 8.33, it follows in particular that G is fsg and that the left invariant measure is unique and also right invariant.

One direction is relatively easy.

PROPOSITION 8.38. *Assume that G is compactly dominated, then it admits a left invariant smooth measure.*

PROOF. Let π_0 denote the canonical projection from G to G/G^{00} and h the Haar measure on the latter. We define a global Keisler measure μ on G as follows: Let D be a $\mathcal{U}$-definable set. Define $U = \{\bar{g} \in G/G^{00} : \pi_0^{-1}(\bar{g}) \subseteq D\}$ and $C = \{\bar{g} \in G/G^{00} : \pi_0^{-1}(\bar{g}) \cap D \neq \emptyset\}$. Then U is open, C is closed and by compact domination, $h(U) = h(C)$. We let $\mu(D)$ be their common value. It is clear that μ is a G-invariant measure.

Fix any model M over which G is defined. We have a canonical continuous map $f : S_G(M) \to G/G^{00}$ and π_0 factorizes through f. In particular, μ is M-invariant and G is compactly dominated by $(S(M), \mu, \text{tp}(\cdot/M))$. By Lemma 8.35, the measure μ is smooth. ⊣

In order to prove the converse, we first explain how to restrict to the case where L is countable. Assume that G admits a left invariant smooth measure μ. Fix a definable set D and we have to show that $\{\bar{g} \in G/G^{00} : \pi_0^{-1}(\bar{g}) \cap D \neq \emptyset$ and $\pi_0^{-1}(\bar{g}) \cap D^c \neq \emptyset\}$ has measure 0. There is a countable sublangage $\tilde{L}$ sufficient to define D and such that the restriction of μ to $\tilde{L}$ is smooth. Denote by $\tilde{G}^{00}$ the connected component of G in the reduct to $\tilde{L}$. Assuming we know the theorem for countable languages, then the set $\tilde{U} = \{\bar{g} \in G/\tilde{G}^{00} : \pi_0^{-1}(\bar{g}) \cap D \neq \emptyset$ and $\pi_0^{-1}(\bar{g}) \cap D^c \neq \emptyset\}$ has measure 0. Let $f : G/G^{00} \to G/\tilde{G}^{00}$ be the canonical map. The Haar measure on $G/\tilde{G}^{00}$ is the pushforward by f of the Haar measure h on G/G^{00}. Set $U = \{\bar{g} \in G/\tilde{G}^{00} : \pi_0^{-1}(\bar{g}) \cap D \neq \emptyset$ and $\pi_0^{-1}(\bar{g}) \cap D^c \neq \emptyset\}$. Then $U = f^{-1}(\tilde{U})$ and $h(U) = 0$ as required.

Assume from now on that L is countable. Let G be a definable fsg group whose invariant measure μ is smooth. Fix a countable model M over which G is defined (and hence μ is M-invariant). If a, b are two elements of $\mathcal{U}$ with $\text{tp}(a/Mb)$ weakly random for μ, it is not always the case that $\text{tp}(b/Ma)$ is weakly random. To remedy this, we introduce a stronger notion of randomness.

Fix a countable elementary submodel $\mathbb{U}$ of the set theoretic universe containing L, T, M, G, μ etc. If $a \in \mathcal{U}$ is a finite tuple, a point $b \in G(\mathcal{U})$ is said to be *random* over Ma if there does not exist some Borel set $B \subseteq S_{xy}(M)$ in $\mathbb{U}$ such that $B(a,b)$ holds and $\mu(B(a,y)) = 0$. Note that such a b always exists because we have to avoid countably many Borel sets of measure 0.

LEMMA 8.39. *If $g \in G$ is random over Mh, then $h \cdot g$ is random over Mh.*

PROOF. This follows from left-invariance of μ: if $\mu(B(x)) = 0$ then also $\mu(B(h \cdot x)) = 0$. ⊣

LEMMA 8.40. *Let a be random over M and b be random over Ma, then a is random over Mb.*

PROOF. The proof uses symmetry of μ. Assume the conclusion does not hold and let $B(x,y)$ in $\mathbb{U}$ be Borel over M such that $B(a,b)$ holds and $B(x;b)$ has measure 0. Then the set of points b' such that $\mu(B(x,b')) = 0$ is Borel over M and lies in $\mathbb{U}$. As it contains b, it must have non-zero measure. Restricting the y variable to that set, we may assume that $\mu(B(x,b')) = 0$ for all b'. But then by symmetry of μ, $\mu^{(2)}(B(x,y)) = \int_{b'} \mu(B(x,b'))d\mu = 0$. Hence the set of points a' such that $\mu(B(a',y)) > 0$ has measure 0 and also lies in $\mathbb{U}$. So a does not lie in that set and $\mu(B(a,y)) = 0$. This contradicts the fact that b is random over Ma. ⊣

PROPOSITION 8.41. (T *is countable.*) *Assume that G admits a left invariant measure $\mu(x)$ which is smooth over M, then G is compactly dominated.*

PROOF. As T is countable, we may assume that M is also countable. Fix a notion of randomness as explained above. Let μ_0 be the restriction of μ to M and let $f : S(\mathcal{U}) \to S(M)$ be the restriction map.

CLAIM. *For all $b \in \mathcal{U}$, for all $g \in G(\mathcal{U})$ random over Mb, for all $g' \in G(\mathcal{U})$, we have*

$$\mathrm{tp}(g/M) = \mathrm{tp}(g'/M) \Longrightarrow \mathrm{tp}(g/Mb) = \mathrm{tp}(g'/Mb).$$

PROOF OF CLAIM. Proof: Let $\phi(x;y) \in L(M)$ and define $B_\phi(x,y) = \{(g,b) \in \mathcal{U} : \mathrm{tp}(g/M) \vdash \phi(x;b) \text{ or } \mathrm{tp}(g/M) \vdash \neg\phi(x;b)\}$. Then $B_\phi(x,y)$ is Borel over M (in fact open) and by compact domination, for any $b' \in \mathcal{U}$, we have $\mu(B(x,b')) = 1$. In particular the Borel set $B(x,b) = \bigcap_{\phi \in L(M)} B_\phi(x,b)$ has μ-measure 1. By the definition of randomness, $\mathrm{tp}(g/Mb)$ does not lie in that set. The claim follows. ⊣ (Claim)

Pick now $g \in G(\mathcal{U})$ random over M and let $g' \in G(\mathcal{U})$ be any element in the same G^{00}-coset as g. We will show that $\mathrm{tp}(g/M) = \mathrm{tp}(g'/M)$.

Let $h \in G(\mathcal{U})$ be random over Mgg'. Then in particular, $\mathrm{tp}(h/Mgg')$ is f-generic. As f-generic types are G^{00}-invariant, we have $\mathrm{tp}(g^{-1} \cdot h/M) = \mathrm{tp}(g'^{-1} \cdot h/M)$. Hence

$$\mathrm{tp}(h^{-1} \cdot g/M) = \mathrm{tp}(h^{-1} \cdot g'/M). \qquad (*)$$

Now as h is random both over Mg and Mg', Lemma 8.40 tells us that g is random over Mh and by Lemma 8.39, $h^{-1} \cdot g$ is random over Mh. Therefore by $(*)$ and the claim, $\mathrm{tp}(h^{-1} \cdot g/Mh) = \mathrm{tp}(h^{-1} \cdot g'/Mh)$. It follows that $\mathrm{tp}(g/M) = \mathrm{tp}(g'/M)$.

Consider the canonical map $f : S(M) \to G/G^{00}$. We have shown that there is some Borel set $X \subseteq S(M)$ of μ_0-measure 1 such that for $p \in X$, $f^{-1}(f(p)) = \{p\}$. Let $X' = f(X)$. Then X' has measure 1 and any point in X' has a unique preimage under f. Let $\pi_1 : S(\mathcal{U}) \to S(M)$ and $\pi_0 : S(\mathcal{U}) \to G/G^{00}$ be the canonical projections. Then $\pi_0 = f \circ \pi_1$. By Lemma 8.35, G is compactly dominated by $(S(M), \mu_0, \pi_1)$ and by the properties of X', it follows that G is compactly dominated by $(G/G^{00}, \pi_0)$. ⊣

Example 8.42. Any definably compact group in an o-minimal structure is compactly dominated, as is the group $(\mathbb{Z}_p, +)$ of p-adic integers. A direct proof was sketched in Example 8.34 for the o-minimal case. More generally in the class of *distal* theories (see Chapter 9), every fsg group is compactly dominated.

References and related subjects

The proof that G^{00} exists for any NIP group is from Shelah [108]. The existence of G^{∞} was proved by Shelah [101] in the abelian case and then generalized by Gismatullin [42] in the general case.

Definably amenable groups were introduced by Hrushovski, Peterzil and Pillay in [60], where the notions of compact domination and fsg group appear. The sequel [61] by Hrushovski and Pillay makes the connection with f-generic types and proves both uniqueness of the invariant measure for fsg groups and compact domination for definably compact groups in o-minimal theories.

Theorem 8.37 comes from unpublished work by Hrushovski, Macpherson and Pillay.

A lot has been written on *stable groups*. Poizat's monograph [96] is an excellent reference, as is Wagner's more technical [121].

In [76] and [84], Newelski suggests studying the action of a group $G(M)$ on the type space $S_G(M)$ as an instance of a dynamical system and investigates the relevance of tools from *topological dynamics* to the model theoretic context. In particular, he asks whether for NIP groups, the so-called *Ellis group* of the aforementioned action coincides with G/G^{00} and proves that it is indeed the case for definably compact groups in o-minimal theories. Gismatullin, Penazzi and Pillay show in [44] that this fails for $SL_2(\mathbb{R})$. The question is still open in the case of a definably amenable group. Positive results are obtained by Pillay in [90] for fsg groups and by Chernikov, Pillay and myself in [26] for groups with a definable f-generic type (in some sense, the opposite extreme).

The first example where G^{00} *does not equal* G^{∞} was given by Conversano and Pillay in [30]. The group in question is the universal cover of $SL_2(\mathbb{R})$. They also give a related example which is interpretable in an o-minimal theory. Techniques to construct groups where the two connected components differ are more systematically explored by Gismatullin and Krupiński in [43].

Chapter 9

DISTALITY

One basic intuition we have about NIP structures is that they are somehow built out of stable components and linear orders. In this view, the theory ACVF of algebraically closed valued fields is archetypical: it has a stable part embodied in the residue field, an order part which is the value group, and the whole structure is in some sense dominated by those two components. Stable theories appear as NIP theories which are degenerate in a certain way. The idea of distality is to characterize the other extreme: NIP theories which are as far away from stable as possible.

A first attempt towards such a definition might be to ask for the absence of non-realized generically stable types. However this is not a well-behaved notion. For example it need not pass from T to T^{eq}. Generically stable types are too rare in general to encode enough information. On the other hand, we have seen that generically stable measures are more widespread. The condition that all generically stable measures are smooth turns out to be much more robust and to have a number of equivalent formulations. Since measures are not convenient to work with, we give the definition in terms of invariant types. We will see in the end that the two conditions are equivalent. Also we show that distal theories enjoy a strong form of honest definitions which is a powerful tool to work with.

Apart from subsection 9.3.2, this chapter does not require familiarity with measures.

Throughout this chapter, we assume that T is NIP.

9.1. Preliminaries and definition

Recall that by an invariant type, we mean a global type invariant over some small set of parameters. Two invariant types $p(x)$ and $q(y)$ *commute* if $p(x) \otimes q(y) = q(y) \otimes p(x)$. If p and q are M-invariant, this is in general stronger than saying that $p(x) \otimes q(y)|_M = q(y) \otimes p(x)|_M$. However, if N contains M and is $|M|^+$-saturated, then $p(x) \otimes q(y)|_N = q(y) \otimes p(x)|_N$

implies that p and q commute (since any formula with parameters in $\mathcal{U}$ has an $\mathrm{Aut}(\mathcal{U}/M)$-conjugate over N).

A notion which is stronger than commuting is orthogonality. Two types $p(x)$ and $q(y)$ over the same set A are *weakly orthogonal* if $p(x) \cup q(y)$ implies a complete type over A. Two global M-invariant types $p(x)$ and $q(y)$ are *orthogonal* if they are weakly orthogonal as global types. Equivalently, if $p|_N$ and $q|_N$ are weakly orthogonal, where $N \succ M$ is $|M|^+$-saturated. Again, it may happen that $p|_M$ and $q|_M$ are weakly orthogonal, but p and q are not orthogonal.

Example 9.1. Any generically stable type commutes with all invariant types, in particular with itself. It is however not orthogonal to itself unless it is a realized type.

Take T to be DLO. Then any two 1-types over a model M are either equal or weakly orthogonal. If p and q are two global invariant 1-types, then they commute if and only if they are orthogonal if and only if they are distinct. In higher dimensions two distinct types need not be orthogonal, but we still have that two invariant types commute if and only if they are orthogonal. As we will see, this latter property is equivalent to distality of the theory.

Exercise 9.2. Give an example in the theory of dense trees of an M-invariant type p which satisfies $p(x) \otimes p(y)|_M = p(y) \otimes p(x)|_M$, but is not generically stable.

We now give the main definition of this chapter.

DEFINITION 9.3. Let p be a global A-invariant type. We say that p is *distal* over A if for any tuple $b \in \mathcal{U}$, if $I \models p^{(\omega)} \restriction Ab$, then the two types $p \restriction AI$ and $\mathrm{tp}(b/AI)$ are weakly orthogonal.

If p is not distal over A, then there is some $b \in \mathcal{U}$, $I \models p^{(\omega)}|_{Ab}$ and $a \models p|_{AI}$ such that a does not realize p over AbI. Such a triple (I, a, b) will be called a *witness of non-distality* of p over A. Note that there is some finite initial segment I_0 of I such that already a does not realize p over AbI_0. Write $I = I_0 + I_1$ and set $b' = I_0 b$. Then (I_1, a, b') is a witness of non-distality with the additional property that a does not realize p over Ab'.

If p is generically stable and distal over some A, then it is realized. To see this take in the definition b to realize $p|_A$, then also $b \models p|_{AI}$ by commutativity, hence the two types $p|_{AI}$ and $\mathrm{tp}(b/AI)$ are equal. Being weakly orthogonal, they must be algebraic, hence p is a realized type.

Example 9.4. In the theory of dense linear orders, any A-invariant type p is distal over A because the sequence I forces the two types $p|_{AI}$ and $\mathrm{tp}(b/AI)$ to fall in different cuts.

Example 9.5. Take T to be DLO expanded by a binary predicate E which defines an equivalence relation with infinitely many dense-co-dense classes. Let

p be the 1-type of an element at $+\infty$ lying in a new equivalence class. Then p is not distal: with the notations as above, one can take $b \models p|_A$ and then $a \models p|_{AI}$ to be in the same E-class as b.

The non-distality of p is due here to the existence of a generically stable type in the imaginary sort of E-classes. We will see later that in fact if there is a non-distal type in T^{eq}, then there is one in T. In Section 9.3.3, we give an example of a non-distal theory with no (non-realized) generically stable types, even in T^{eq}.

9.2. Base change lemmas

The main technical difficulty in the study of distality has to do with base changes: we are given some construction over a base A and we want to carry out a similar construction over a larger set B. The following lemmas will enable us to do such base changes when dealing with witnesses of non-distality.

LEMMA 9.6. *Let p be a global A-invariant type and $b \in \mathcal{U}$. Let $I \models p^{(\omega)}|_{Ab}$ and $a \models p|_{AI}$. Let $B \supseteq A$. Then there are a', I', b' such that $I' \models p^{(\omega)}|_{Bb'}$, $a' \models p|_{BI'}$ and* $\text{tp}(a'b'/A) = \text{tp}(ab/A)$.

Furthermore, if $b \models q|_A$ for some A-invariant type q which commutes with p, then we may add the condition $b' \models q|_{BI'}$.

PROOF. First observe that for any linear order $\mathfrak{J}$, we can find J such that J is a Morley sequence of p over Ab of order type $\mathfrak{J}$ and $a \models p|_{AJ}$. Indeed, it suffices to construct a Morley sequence of $\lim(I)$ over $AIab$ indexed by the opposite order $\mathfrak{J}^*$ and take for J that sequence read in the reverse direction.

Assume the conclusion is false and let $I' \models p^{(\omega)}|_{Bb}$. Then by compactness, there is a formula $\theta(x) \in p|_{BI'}$ such that whenever a', b' satisfy $\text{tp}(a'b'/A) = \text{tp}(ab/A)$ and $I' \models p^{(\omega)}|_{Bb'}$, then $\models \neg\theta(a')$.

Set $b_0 = b$ and $I_0 = I'$. As $I_0 b_0 \equiv_A Ib$, there is $a_0 \in \mathcal{U}$ such that $I_0 b_0 a_0 \equiv_A Iba$. By the previous paragraph, we must have $\models \neg\theta(a_0)$. Now let $I_1 \models p^{(\omega)} \restriction BI_0a_0$. Then $I_0 + I_1$ is a Morley sequence of p over B and $I_0 + (a_0) + I_1$ is indiscernible over A. By the first paragraph of the proof, we may find $a_1, b_1 \in \mathcal{U}$ such that $a_1 b_1 \equiv_A ab$, $I_0 + (a_0) + I_1$ is a Morley sequence of p over Ab_1 and $I_0 + (a_0) + I_1 + (a_1)$ is indiscernible over A. As previously, we have $\neg\theta(a_1)$.

Proceed by taking $I_2 \models p^{(\omega)} \restriction BI_0I_1a_0a_1$ and iterate this construction. After ω steps we obtain a sequence $I_* = I_0 + (a_0) + I_1 + (a_1) + I_2 + (a_2) + \cdots$ which is a Morley sequence of p over A, such that $I_0 + I_1 + \cdots$ is a Morley sequence of p over B and $\models \neg\theta(a_i)$ for each $i < \omega$. But then the formula $\theta(x)$ alternates infinitely often on the sequence I_*, contradicting NIP.

For the 'furthermore' part, first note that by the commutativity assumption, for any linear order $\mathfrak{I}$, if $I' \models p^{(\mathfrak{I})}|_B$ and $b' \models q|_B$, then $I' \models p^{(\mathfrak{I})}|_{Bb'}$ if and only if $b' \models q|_{BI'}$. Keeping this in mind, the proof is the same except that at each step, we take b_k realizing q over the union of B and the Morley sequence of p constructed so far: At the first step, we pick $b_0 \models q|_B$; at the second step, we take $b_1 \models q|_{BI_0I_1a_0}$ then find a_1 satisfying the requirements, and so on. The contradiction at the end is obtained in the exact same way. ⊣

PROPOSITION 9.7. *If p is a global type invariant over both A and B, then p is distal over A if and only if it is distal over B.*

PROOF. We may assume that $A \subseteq B$. First suppose that p is distal over A. Let $b \in \mathcal{U}$ and $I \models p^{(\omega)}|_{Bb}$. Then setting $b' = Bb$, we have $I \models p^{(\omega)}|_{Ab'}$, hence by distality over A, $p|_{AI}$ and $\text{tp}(b'/AI)$ are weakly orthogonal. A fortiori so are $p|_{BI}$ and $\text{tp}(b/BI)$.

Conversely, assume that p is not distal over A. There are $b \in \mathcal{U}$, $I \models p^{(\omega)}|_{Ab}$ and $a \models p|_{AI}$ such that $a \not\models p|_{Ab}$. By Lemma 9.6, we can find a', b', I' such that $I' \models p^{(\omega)}|_{Bb'}$, $a' \models p|_{BI'}$ and $\text{tp}(a'b'/A) = \text{tp}(ab/A)$. Then $a' \not\models p|_{Bb'}$ and p is not distal over B. ⊣

If p is a global invariant type, we will say that p is distal if it is distal over some (equiv. any) A over which it is invariant.

LEMMA 9.8. *Let p be a global A-invariant type and $I = (b_i : i \in \mathfrak{I})$ an indiscernible sequence where $\mathfrak{I}$ has cofinality $> 2^{|T|+|A|}$. Let $q = \lim(I)$ be the limit type of I. Then p and q commute.*

PROOF. Let $m \in \mathcal{U}$ and $\phi(x, y; m)$ such that $p(x) \otimes q(y) \vdash \phi(x, y; m)$. Let $b \models q|_{AIm}$. By the cofinality assumption, there is some $i_m \in \mathfrak{I}$ such that $\text{tp}(b_i/Am)$ is constant for $i \geq i_m$. As b realizes the limit type of I over AIm, we have $\text{tp}(b/Am) = \text{tp}(b_i/Am)$ for any $i \geq i_m$. Let $a \models p|_{AImb}$. By assumption we have $\models \phi(a, b; m)$. Then by A-invariance of p, also $\phi(a, b_i; m)$ holds for $i \geq i_m$. Therefore if $b' \models q|_{AIma}$, then $\phi(a, b'; m)$ holds and this shows that $q(y) \otimes p(x) \models \phi(x, y; m)$ as required. ⊣

The hypothesis that $\mathfrak{I}$ has large cofinality can be removed if we assume that I is A-indiscernible; see Exercise 9.16.

LEMMA 9.9. *Let p be A-invariant. Let $b \in \mathcal{U}$, $I \models p^{(\omega)}|_{Ab}$ and $a \models p|_{AI}$. Then there is some set $B \supseteq A$, a global B-invariant type q and a', b', I' such that:*

- *q commutes with p;*
- *$I' \models p^{(\omega)}|_{Bb'}$ and $a' \models p|_{BI'}$;*
- *$b' \models q|_{BI'}$;*
- *$\text{tp}(a'b'/A) = \text{tp}(ab/A)$.*

PROOF. Let $\kappa = (2^{|T|+|A|})^+$ and $\lambda = \beth_\kappa$. Construct inductively a sequence $J = (I_k, a_k, b_k : k < \lambda)$ such that $(I_0, a_0, b_0) = (I, a, b)$ and for each $0 < k < \lambda$, (I_k, a_k, b_k) is given by Lemma 9.6 with $B = AI_{<k}a_{<k}b_{<k}$. By Erdős-Rado (Proposition 1.1), there is an A-indiscernible sequence $(I'_i, a'_i, b'_i : i < \kappa + 1)$

such that for any n, there are $i_0 < \cdots < i_{n-1} < \lambda$ with

$$I_{i_0} a_{i_0} b_{i_0} \ldots I_{i_{n-1}} a_{i_{n-1}} b_{i_{n-1}} \equiv_A I'_0 a'_0 b'_0 \ldots I'_{n-1} a'_{n-1} b'_{n-1}. \qquad (*)$$

Set $B = A \cup \{b'_i : i < \kappa\}$. Then $(*)$ ensures that $I'_\kappa \models p^{(\omega)}|_{Bb'_k}$, $a'_k \models p|_{BI'_k}$ and $\mathrm{tp}(a'_\kappa b'_\kappa/A) = \mathrm{tp}(ab/A)$. Define q as the limit type of the sequence $(b'_i : i < \kappa)$. By Lemma 9.8, p and q commute. Also $b'_\kappa \models q|_{BI'_\kappa}$ since the sequence $(b'_i : i \leq \kappa)$ is indiscernible over I'_κ. Hence we have what we want if we set $(I', a', b') = (I'_\kappa, a'_\kappa, b'_\kappa)$. ⊣

LEMMA 9.10. *Let p be a global invariant type. Assume that p is not distal, then there are $M \prec N$ such that N is $|M|^+$-saturated, p is invariant over M and there is (I, a, b), a witness of non-distality of p over N such that for some M-invariant type q, $b \models q|_{NI}$ and q commutes with p.*

We may furthermore assume that a does not realize p over Mb.

PROOF. We just have to apply one after the other some of the previous lemmas. Let $A \subset \mathcal{U}$ be such that p is A-invariant. Then there is (I, a, b) a witness of non-distality of p over A. By incorporating a finite initial segment of I into b, we may assume that $\mathrm{tp}(a/Ab)$ is not equal to $p|_{Ab}$. Then apply Lemma 9.9 to obtain some set B, sequence I' and tuples a', b' giving a witness of non-distality of p over B and such that $b' \models q|_{BI'}$, where q is a global B-invariant type commuting with p. Let N be a $|B|^+$-saturated model containing B. We apply again Lemma 9.6 to obtain (I'', a'', b'') a witness of non-distality of p over N such that $b'' \models q|_{NI''}$.

Furthermore, the construction ensures that $\mathrm{tp}(a'', b''/A) = \mathrm{tp}(a', b'/A) = \mathrm{tp}(a, b/A)$. As a does not realize p over Ab, this gives the second part of the lemma. ⊣

PROPOSITION 9.11. *Let p be a global invariant type, then p is distal if and only if p is orthogonal to every global invariant type to which it commutes.*

PROOF. First assume that p is distal and let q be a global invariant type which commutes with p. Let M be a small model over which both p and q are invariant. Let $I \models p^{(\omega)}|_M$. We show that $p(x)|_{MI} \cup q(y)|_{MI}$ implies a complete type in variables $x\hat{\ }y$ over M. As M can be taken as large as we want, this shows that p and q are orthogonal. Take any $b \models q|_{MI}$. As q commutes with p, we also have $I \models p^{(\omega)}|_{Mb}$. Hence by distality of p over M, $p|_{MI}$ and $\mathrm{tp}(b/MI)$ are weakly orthogonal.

Conversely, assume that p is not distal. Let $M \prec N$ and q as in the previous lemma. Then the types $p|_N$ and $q|_N$ are not weakly orthogonal. Since both are invariant over M and N is $|M|^+$-saturated, p and q are not orthogonal. ⊣

Exercise 9.12. Let $p(x)$ be a distal A-invariant type and $f(x)$ an A-definable function. Then the pushforward $f_*(p)$ defined by $f_*(p) \vdash \phi(y)$ if and only if $p \vdash \phi(f(x))$ is also A-invariant and distal. In particular any non-distal type in T^{eq} lifts to a non-distal type in T.

[Observe first that the proof of Lemma 9.10 gives a type q which commutes with all invariant types based on A.]

PROPOSITION 9.13. *Let p be a global invariant type of dp-rank 1. Then p is either generically stable or distal.*

PROOF. Take $A \subset \mathcal{U}$ such that p is A-invariant and $p|_A$ already has dp-rank 1. If p is not distal, then there is a global invariant type q which commutes with p and such that p and q are not orthogonal. Increasing A if necessary, we may assume that q is also A-invariant. Let $I \models p^{(\omega)}|_A$, $J \models q^{(\omega)}|_{AI}$ and $B = AIJ$. By non-orthogonality of p and q, there are $a, b \in \mathcal{U}$ such that $a \models p|_B$, $b \models p|_B$ and a does not realize p over Bb. Let $I' \models p^{(\omega)}|_{Bab}$ and $J' \models q^{(\omega)}|_{BI'ab}$. By commutativity of p and q, the two sequences $I + I'$ and $J + (b) + J'$ are mutually indiscernible over A (Example 4.5). We know that $J + (b) + J'$ is not indiscernible over Aa. Hence as dp-rk$(a/A) = 1$, the sequence $I + I'$ must be indiscernible over a. But $I + (a) + I'$ is a Morley sequence of p over A, so this can only happen if that latter sequence is totally indiscernible, that is if p is generically stable. ⊣

Exercise 9.14. Prove Lemma 9.6 without assuming NIP, but assuming instead that p (and q for the 'furthermore' part) is invariant over some $C \subset A$ and A is a $|C|^+$-saturated model.

Exercise 9.15. Let p be an A-invariant type. Show that if p is not distal, then there is a witness of non-distality (I, a, b) over A with $b \models p^{(n)}|_A$ for some n.

Exercise 9.16. Prove that if p is an A-invariant type and I is any A-indiscernible sequence, then p and $\lim(I)$ commute.

9.3. Distal theories

DEFINITION 9.17. We say that the NIP theory T is *distal* if all invariant types are distal.

Note that if T is distal, then so is T^{eq} by Exercise 9.12.

THEOREM 9.18. *Assume that every invariant type in one variable is distal, then T is distal.*

PROOF. Assume that every invariant type in one variable is distal and let p be any global type invariant over some set A. We can find (I, a, b) a witness of non-distality of p over A. Write the tuple b as $b = (b_0, \dots, b_{n-1})$ and assume that (A, I, a, b) was chosen so as to minimize n. Then a realizes p over $AIb_1 \dots b_{n-1}$. Set $A' = A \cup \{b_1, \dots, b_{n-1}\}$, then (I, a, b_0) is a witness of non-distality of p over A'. Hence by minimality of n, we have $n = 1$ and b is a singleton. Incorporating some initial segment of I in the base A, we may assume that a does not realize p over Ab.

Following the same procedure as in the proof of Lemma 9.10, we can find $B \subseteq N$ where N is $|B|^+$-saturated, some B-invariant type q commuting with p and a witness (I', a', b') of non-distality of p over B such that $b' \models q|_B$ and $\text{tp}(a', b'/A) = \text{tp}(a, b/A)$. In particular b' is a singleton, hence by hypothesis q is distal. But then $q|_N$ is weakly orthogonal to $p|_N$ contradicting the fact that a' does not realize p over Ab'. ⊣

COROLLARY 9.19. *If T is dp-minimal with no non-realized generically stable type then it is distal.*

PROOF. If T is dp-minimal, then by Proposition 9.13 every 1-type is either distal or generically stable. If the latter case does not occur, then all 1-types are distal which by the previous theorem implies that T is distal. ⊣

Note the following consequence of this corollary: If T is dp-minimal and has no non-realized generically stable type, then also T^{eq} has no non-realized generically stable type.

Example 9.20. Any dp-minimal linearly ordered theory—in particular any o-minimal theory or weakly o-minimal theory—is distal, as is the theory $Th(\mathbb{Q}_p)$ of p-adics.

The distality of a theory can be stated in a few equivalent ways. One is that every invariant type is distal, equivalently any two invariant types which commute are orthogonal. We will see now two other characterizations in terms of strong honest definitions and generically stable measures.

9.3.1. Strong honest definitions. Perhaps the property of distal theories which is the easiest to use is the following strong form of honest definitions.

Let $\phi(x; y) \in L$ and define as usual $\phi^{opp}(y; x) = \phi(x; y)$. In what follows, we use the notation $\theta(y; d) \vdash \text{tp}_{\phi^{opp}}(b/A)$ to mean that for every $\phi(a; y)^\varepsilon \in \text{tp}_{\phi^{opp}}(b/A)$, $\varepsilon \in \{0, 1\}$, we have $\theta(y; d) \vdash \phi(a; y)^\varepsilon$.

THEOREM 9.21. *Let T be a distal theory. Let $M \models T$, $A \subseteq M$, $\phi(x; y) \in L$ and $b \in M$ a $|y|$-tuple. Then there is an elementary extension $(M, A) \prec (M', A')$, a formula $\theta(y; z) \in L$ and a tuple d of elements of A' such that $M' \models \theta(b; d)$ and $\theta(y; d) \vdash \text{tp}_{\phi^{opp}}(b/A)$.*

If we set $\psi(x; z) = \forall y(\theta(y; z) \rightarrow \phi(x; y))$, then we have $\phi(A; b) = \psi(A; d)$ and $M' \models \psi(x; d) \rightarrow \phi(x; b)$. In particular, $\psi(x; d)$ is an honest definition of $\phi(x; b)$ over A. We can also define $\psi'(x; z) = \forall y(\theta(y; z) \rightarrow \neg\phi(x; y))$. Then the same two properties hold with $\neg\phi$ instead of ϕ. Conversely, if we have some $\psi(x; z)$ and $\psi'(x; z)$ with these properties, than one can recover a formula $\theta(y; z)$ by setting $\theta(y; z) = \forall x(\psi(x; z) \rightarrow \phi(x; y) \wedge \psi'(x; z) \rightarrow \neg\phi(x; y))$.

PROOF. Let $S_A \subset S_x(\mathcal{U})$ be the set of global types finitely satisfiable in A: a compact set. Let $p \in S_A$. As p is finitely satisfiable in A, for any small set

$B \subset A'$, $p(x)|_{MB} \cup \boldsymbol{P}(x)$ is finitely satisfiable. Hence by saturation, there is $a \in A'$ realizing p over MB. In particular, we can find I in A', a Morley sequence of p over M. By distality of p, the types $p|_{AI}$ and $\text{tp}(b/AI)$ are weakly orthogonal. By compactness, there is $\psi_p(x) \in p|_{AI}$ and $\theta_p(y) \in \text{tp}(b/IA)$ such that $\psi_p(x) \wedge \theta_p(y) \vdash \phi(x; y)^{\varepsilon_p}$, where ε_p is such that $p \vdash \phi(x; b)^{\varepsilon_p}$. As S_A is compact, we can find a finite set $S_0 \subset S_A$ such that $\bigcup_{p \in S_0} \psi_p(x)$ covers S_A. Set $\theta(y) = \bigwedge_{p \in S_0} \theta_p(y)$. Write $\theta(y) = \theta(y; d)$ making the parameters explicit.

We check that $\theta(y; d) \vdash \text{tp}_{\phi^{opp}}(b/A)$. Let $a \in A$ and ε such that $\models \phi(a; b)^{\varepsilon}$. There is $p \in S_0$ such that $a \models \psi_p(x)$. Since $\theta_p(y) \wedge \psi_p(x) \to \phi(x; y)^{\varepsilon_p}$ and $\theta_p(b)$ is true, we must have $\varepsilon_p = \varepsilon$. Thus $\theta(y; d)$ implies $\theta_p(y)$ which implies $\phi(a; y)^{\varepsilon}$. ⊣

As for usual honest definitions, there is an equivalent formulation avoiding the use of pairs: Let T be distal, M, A, $\phi(x; y)$ and b as above. Then there is a formula $\theta(y; z)$ such that for every finite $A_0 \subseteq A$, for some $d \in A$ we have $M \models \theta(b; d)$ and $\theta(y; d) \vdash \text{tp}_{\phi^{opp}}(b/A_0)$. Equivalently, there is a formula $\psi(x; z)$ such that for every finite $A_0 \subseteq A$, for some $d \in A$, we have $\psi(A_0; d) = \phi(A_0; b)$ and $M \models \psi(x; d) \to \phi(x; b)$.

Uniformity for this strong form of honest definitions can be proved the same way as for the weak form.

THEOREM 9.22. *Let T be a distal theory and let $\phi(x; y) \in L$. Then there is a formula $\theta(y; z) \in L$ such that for all $M \models T$, $A \subseteq M$ of size at least 2, and $b \in M^{|y|}$, there is an elementary extension $(M, A) \prec (M', A')$ and $d \in A'$ such that $M' \models \theta(b; d)$ and $\theta(y; d) \vdash \text{tp}_{\phi^{opp}}(b/A)$.*

PROOF. We use the formulation with $\psi(x; z)$ as stated above. The proof is then exactly the same as for usual honest definitions: First, as in Lemma 6.15 we use compactness to show that if we associate to each $\psi(x; z)$ an integer q_ψ, there are finitely many $\psi_0, \ldots, \psi_{n-1}$ such that for any pair (M, A) and $\phi(x; b) \in L(M)$, for some $j < n$ and for every $A_0 \subseteq A$ of size $\leq q_{\psi_j}$ there is some $d \in A$ with $\phi(A_0; b) \subseteq \psi_j(M; d) \subseteq \phi(M; b)$. Next we use the (p, q)-theorem as in Theorem 6.16 to conclude. ⊣

COROLLARY 9.23. *Let T be distal and let $\phi(x; y) \in L$. Then there is a formula $\theta(y; z) \in L$ such that for all $M \models T$, $A \subseteq M$ a finite set of size at least 2 and $b \in M^{|y|}$, there is some $d \in A$ such that $\theta(b; d)$ holds and $\theta(y; d) \vdash \text{tp}_{\phi^{opp}}(b/A_0)$.*

Exercise 9.24. Show that conversely if T satisfies the conclusion of Theorem 9.21, then it is distal.

Remark 9.25. It is important to note that distality is not preserved under reducts (since the theory of equality is not distal). This is also witnessed by the fact that in Theorem 9.21 one might have to look for $\theta(y; z)$ outside of the language generated by $\phi(x; y)$. For example if $\phi(x; y)$ is $x = y$, then no formula in the language of equality will work for θ.

In that respect, distality is of a different nature than combinatorial conditions such as stable, NIP, NTP_2 etc.

9.3.2. Generically stable measures. We show here a characterization of distality in terms of generically stable measures: an NIP theory T is distal if and only if all generically stable measures are smooth.

PROPOSITION 9.26. *If T is distal, then all generically stable measures are smooth.*

PROOF. Assume that T is distal and let $\mu(x)$ be a global generically stable measure, invariant over some M. We show that $\mu|_M$ is smooth. Let $M \prec N$, $\phi(x; b) \in L(N)$ and $\varepsilon > 0$. Let (N', M') be a sufficiently saturated extension of the pair (N, M). By Theorem 9.21, there is some $\theta(y; d) \in L(M')$ such that $\theta(b; d)$ holds and $\theta(y; d) \vdash \mathrm{tp}_{\phi^{opp}}(b/M)$. Let $\psi(x; z) = \forall y(\theta(y; z) \to \phi(x; y))$. Then we have $\phi(M; b) = \psi(M; d)$ and $N' \models \psi(x; d) \to \phi(x; b)$. As μ is finitely satisfiable in M, $\mu(\phi(x; b) \Delta \psi(x; d)) = 0$. Furthermore, as μ is definable over M, there is some $\chi(z) \in \mathrm{tp}(d/Mb)$ such that for any $e \models \chi(z)$, $\models \psi(x; e) \to \phi(x; b)$ and $\mu(\phi(x; b)) - \mu(\psi(x; e)) < \varepsilon$. By elementarity of the extension $(N, M) \prec (N', M')$, we can find such a e in M. Applying the same argument for $\neg\phi(x; b)$, we find some $\psi'(x; e') \in L(M)$ such that $\models \phi(x; b) \to \psi'(x; e')$ and $\mu(\psi'(x; e')) - \mu(\phi(x; b)) < \varepsilon$.

Hence $\phi(x; b)$ is sandwiched between $\psi(x; e)$ and $\psi'(x; e')$ and the difference of the μ measures of those two sets is at most 2ε. As ε was arbitrary, this shows that the measure of $\phi(x; b)$ is determined by $\mu|_M$. Therefore μ is smooth. ⊣

PROPOSITION 9.27. *If all generically stable measures are smooth, then T is distal.*

PROOF. Assume that T is not distal and let p be a non-distal invariant type. By Lemma 9.10, we can find $M \prec N$ and some M-invariant type q such that N is $|M|^+$-saturated, p is M-invariant, p and q commute and there is (I, a, b) a witness of non-distality of p over N such that $b \models q|_{NI}$. We may furthermore assume that a does not realize p over Mb. Let $\phi(x; b) \in p|_{Mb}$ such that $\models \neg\phi(a; b)$.

Set $a_0 = a$. Let $\mathfrak{I} = \mathbb{Q} \cap (-1/2, 0)$ with the usual order and $\mathfrak{I}^*$ is $\mathfrak{I}$ with the opposite order. Build a Morley sequence of $\lim(I)$ over everything considered so far with order type $\mathfrak{I}^*$. Let $I_0 = (a_t : t \in \mathfrak{I})$ be that sequence read in the opposite order. Set $\mathfrak{J} = \mathbb{Q} \cap (0, 1/2)$ and let $J_0 = (a_t : t \in \mathfrak{J})$ be a Morley sequence of p over everything. Then:

- the sequence $I_0 + J_0$ is a Morley sequence of p over Nb;
- the sequence $I_0 + (a_0) + J_0$ is a Morley sequence of p over N;
- b realizes q over $N \cup \{a_t : t \neq 0\}$.

The last bullet follows from the first and commutativity of p and q.

Let P be a model containing $I_0 + J_0$ and such that $b \models q|_P$. Let $\mu(x)$ be the average measure of the sequence $I_0 + J_0$ (as in Example 7.2). As explained in Example 7.32, μ is generically stable. We will show that μ is not smooth over P and in fact that $\mu|_P$ has an extension to Pb in which $\mu(\neg\phi(x; b)) > 0$.

If this is not the case, then by Lemma 7.4, there is some formula $\psi(x) \in L(P)$ such that $\models \psi(x) \to \phi(x; b)$ and $\mu(\psi(x)) > 1/2$. Without loss, assume that any a_i in the sequence I_0 satisfies $\psi(x)$.

Fix an arbitrary sequence $(t_i : i < \omega)$ of distinct elements of $(-1/2, 0) \setminus \mathbb{Q}$. We construct inductively points $(a_{t_i} : i < \omega)$ and $(b_i : i < \omega)$ as follows:

Assume that $(a_{t_j} : j < i)$ have been defined. Set $r = t_i$ and let $I_r = (a_t : t < r)$ and $J_r = (a_t : t > r)$, where we include all the a_t constructed so far. Let also b_i realize q over everything.

Then the triple (I_r, J_r, b_i) has the same type over N as (I_0, J_0, b). Hence there is some a_r such that $I_r + (a_r) + J_r$ is a Morley sequence of p over N and $\neg\phi(a_r, b_i)$ holds. As b_i and b both realize q over P, we have $\models \psi(x) \to \phi(x; b_i)$ and therefore $\neg\psi(a_r)$ holds.

Once the construction is finished, we have that $\psi(a_t)$ holds if t is in $\mathbb{Q} \cap (-1/2, 0)$ and $\neg\psi(a_{t_i})$ holds for $i < \omega$. This implies that the formula $\psi(x)$ has infinite alternation rank, contradicting NIP. ⊣

9.3.3. Indiscernible sequences. The first definition of distality in [114] was given in terms of indiscernible sequences. They are no longer required in the approach presented here. However for completeness, we give the definition of distal indiscernible sequences and point out the links with previous notions. We will leave the proofs as exercises to the reader.

In the following definition, I can be an arbitrary sequence, although we are mainly interested in the case where I is indiscernible.

DEFINITION 9.28. Let I be any infinite sequence of tuples and A a base set of parameters. We say that I is distal over A if whenever J, a, $B \supseteq A$ satisfy:

- J is indiscernible and realizes the EM-type of I;
- $J = J_1 + (a) + J_2$, where both J_1 and J_2 are infinite with no endpoints;
- the sequence $J_1 + J_2$ is indiscernible over B;

then the sequence $J = J_1 + (a) + J_2$ is indiscernible over B.

PROPOSITION 9.29. *An A-invariant type p is distal over A if and only if its Morley sequence is distal over A.*

An A-indiscernible sequence I is distal over A if and only if its limit type $\lim(I)$ *is distal.*

PROOF. The implication: if p is not distal over A then its Morley sequence is not distal over A was observed (and used) in the proof of Proposition 9.27. We leave the other verifications to the reader. ⊣

In particular, if I is indiscernible both over A and B, then it is distal over A if and only if it is distal over B. It makes sense therefore to speak of a distal indiscernible sequence without mentioning the set A.

PROPOSITION 9.30. *Let $I = (a_i : i \in [0, 1])$ be an indiscernible sequence and $\mu(x)$ be the average measure of I as in Example 7.2. Then μ is smooth if and only if I is distal.*

PROOF. Left to the reader (again, one direction is essentially included in the proof of Proposition 9.27). ⊣

9.3.4. An example. We end this chapter with an example of a non-distal theory with no non-trivial generically stable type, even in T^{eq}.

The language is $L = \{\leq, R_n : n < \omega\}$ where all are binary predicates. The theory T states that the reduct to $L_0 = \{R_n : n < \omega\}$ is a local order, as presented in Example 2.31: it is the theory of $(\mathbb{Q}; R_n : n < \omega)$ where $R_n(x; y)$ holds if and only if $x < y \wedge |y - x| \leq n$. In addition, we impose that $\leq$ defines a dense linear order with no endpoint, which is generic with respect to the R_n's in the sense that any infinite set which is L_0-definable (with parameters) is dense co-dense with respect to $\leq$.

A straightforward back-and-forth argument shows that this defines a consistent complete theory with elimination of quantifiers. As the structure is linearly ordered, there are no (non-trivial) generically stable types over the real sort. Furthermore, there are no definable equivalence relations with infinite classes, hence no generically stable types in imaginary sorts either (since if $b \in \text{acl}(a)$ and $\text{tp}(a/M)$ is generically stable, then so is $\text{tp}(b/M)$).

It remains to check that T is not distal. Let M be a model of T and let $p_0(x)$ be the unique non-realized L_0-type in one variable over $\mathcal{U}$ which is generically stable. Namely: $p_0 \vdash \neg R_n(x; a)$ for every $a \in \mathcal{U}$. Let p be any expansion of p_0 to an L-type over $\mathcal{U}$ invariant over some small M. Take $b \models p|_M$ and $I \models p^{(\omega)}|_{Mb}$. Then b is R_n-incomparable with any element of M or I. One can find a point $a \models p|_{MI}$ which is R_1-comparable to b. This shows that $p|_{MI}$ and $\text{tp}(b/MI)$ are not weakly-orthogonal. Hence p is not distal.

References and related subjects

Distal theories were introduced by the author in [114]. The approach there is focussed on indiscernible sequences. The version we present here has been reworked entirely. We believe that the definition we start with, of a distal invariant type, is more natural and easier to work with.

The word *distal* comes from dynamical systems. If a group G acts on a (compact or metric) space X, then two points of X are distal if they do not become arbitrarily close when the group acts. There is no mathematical relationship between the two notions. The idea behind using the word here is that in distal theories, two points lying at different places in an indiscernible sequences cannot interact and we like to think of them being separated, or distant, from each other.

Strong honest definitions come from Chernikov and Simon [27].

When a NIP theory is not distal, one can define a non-trivial notion of independence called *s-independence* which is symmetric and has bounded weight. This is done in the second part of [114]. If p and q are two commuting types over a sufficiently saturated model M, then $a \models p$ and $b \models q$ are s-independent over M if and only if $\mathrm{tp}(a, b/M) = p \otimes q$. The theory of s-independence shows that in some sense p and q behave with respect to each other as do types in a stable theory.

The theory can also be developed over an indiscernible sequence. The main result is the following theorem which is a weaker form of distality and holds in any NIP theory.

THEOREM 9.31 ([114], Theorem 3.30). *Let p be an A-invariant type and $b \in \mathcal{U}$. Let $I \models p^{(\omega)} \restriction Ab$ and let $J \models p^{(\omega)} \restriction AI$. Then the limit type* $\lim(J/Ab)$ *coincides with $p|_{Ab}$.*

Appendix A

EXAMPLES OF NIP STRUCTURES

In this appendix we present the classical examples of (unstable) NIP theories. We start with linear orders and trees, which are ubiquitous in NIP structures. Closely related to them are o-minimal and C-minimal structures. We then discuss algebraic examples: ordered abelian groups and most importantly valued fields which are often considered as the archetypical NIP structure.

It turns out that most of those structures are dp-minimal (Definition 4.27). As the reader might want to go through this appendix before having read the material in Chapter 4, we recall the definition: A theory T is dp-minimal if for any singleton a and any two infinite sequences of tuples I_0, I_1, if I_k is indiscernible over I_{1-k} for $k = 0, 1$, then there is $k \in \{0, 1\}$ such that I_k is indiscernible over a. This property implies NIP.

A.1. Linear orders and trees

A.1.1. Linear orders. A *colored order* is a structure $(M; <, (C_i)_{i<\alpha})$ where the C_i's are arbitrary unary predicates and where $<$ defines a linear order on M.

Lemma A.1 (Rubin). *Let $(M; <, (C_i)_{i<\alpha})$ be a colored order and let $a \in M$. Let $\bar{b}_1 \in M$ be a tuple of points all smaller than a and $\bar{b}_2 \in M$ a tuple of points greater than a, then* $\mathrm{tp}(\bar{b}_1/a) \cup \mathrm{tp}(\bar{b}_2/a) \vdash \mathrm{tp}(\bar{b}_1, \bar{b}_2/a)$.

Proof. The proof is a straightforward back-and-forth.

Assume M is ω-saturated and we are given two pairs of tuples $(\bar{b}_1, \bar{b}_2)$ and $(\bar{b}'_1, \bar{b}'_2)$ satisfying the hypothesis and such that $\mathrm{tp}(\bar{b}_1/a) = \mathrm{tp}(\bar{b}'_1/a)$ and $\mathrm{tp}(\bar{b}_2/a) = \mathrm{tp}(\bar{b}'_2/a)$. Notice that necessarily $\mathrm{qftp}(\bar{b}_1, \bar{b}_2, a) = \mathrm{qftp}(\bar{b}'_1, \bar{b}'_2, a)$. Assume we are given $b \in M$, without loss, $b < a$. We can find $b' \in M$ such that $\mathrm{tp}(\bar{b}_1{}^\frown b/a) = \mathrm{tp}(\bar{b}'_1{}^\frown b'/a)$. Thus the back-and-forth goes through. ⊣

Proposition A.2. *Any colored order $(M; <, (C_i)_{i<\alpha})$ is dp-minimal (in particular NIP).*

PROOF. Let $a \in M$ be a singleton and $I = (\bar{a}_i : i \in \mathbb{Q})$ be an indiscernible sequence of n-tuples. For each index i write $\bar{a}_i = (a_i^1, \dots, a_i^n)$. We say that I is *cut* by a if there is $k \leq n$ and $i, i' \in \mathbb{Q}$ such that $a_i^k \leq a \leq a_{i'}^k$.

Assume that I is not cut by a, then it follows from Lemma A.1 that for any $i < i' \in \mathbb{Q}$, $\mathrm{tp}(a/\bar{a}_i{}^\frown\bar{a}_{i'}) \cup \mathrm{tp}((\bar{a}_j : i < j < i')/\bar{a}_i{}^\frown\bar{a}_{i'}) \vdash \mathrm{tp}(a + (\bar{a}_j : i < j < i')/\bar{a}_i{}^\frown\bar{a}_{i'})$. As $(\bar{a}_j : i < j < i')$ is indiscernible over $\bar{a}_i{}^\frown\bar{a}_{i'}$, it follows that $(\bar{a}_j : i < j < i')$ is indiscernible over a and therefore I is indiscernible over a.

If now I and J are two mutually indiscernible sequences, then a singleton a cannot cut both I and J (otherwise either some point of I would cut J or some point in J would cut I). Therefore at least one of I and J is indiscernible over a. This proves dp-minimality. ⊣

A.1.2. Trees. Trees were already introduced in Section 2.3.1 where we studied dense trees. Nevertheless, we recall here all the relevant definitions.

DEFINITION A.3 (Tree). A *tree* is a partially ordered set $(M, \leq)$ such that for every $a \in M$, the set $\{x \in M : x \leq a\}$ is linearly ordered by $\leq$ and for any $a, b \in M$, there is some c smaller or equal to both a and b.

We say that $(M, \leq)$ is a *meet-tree* if in addition: for every two points $a, b \in M$, the set $\{x \in M : x \leq a \wedge x \leq b\}$ has a greatest element, which we denote by $a \wedge b$.

Let $(M, \leq)$ be a meet-tree and $c \in M$ is a point. The *closed cone* of center c is by definition the set $C(c) := \{x \in M : x \geq c\}$. We can define on $C(c)$ a relation E_c by: xE_cy if $x \wedge y > c$. One can easily check that this is an equivalence relation. We define an *open cone* of center c to be a equivalence class under the relation E_c.

A *leaf* of the tree M is a point in M which is maximal.

LEMMA A.4. *Let $(M, \leq)$ be a tree, $a \in M$, and let C denote the closed cone of center a. Let $\bar{x} = (x^1, \dots, x^n) \in (M \setminus C)^n$ and $\bar{y} = (y^1, \dots, y^m) \in C^m$. Then $tp(\bar{x}/a) \cup tp(\bar{y}/a) \vdash tp(\bar{x} \cup \bar{y}/a)$.*

PROOF. The proof is a straightforward back-and-forth as for Lemma A.1. ⊣

PROPOSITION A.5. *Any tree $(M, \leq)$ is dp-minimal (hence NIP).*

The proof is similar to that of Proposition A.2, but slightly longer as we have various cases to consider. We omit it here and refer the reader to [112, Proposition 4.7].

A.1.3. O-minimality. Let $(M; <, \dots)$ be a structure on which $<$ defines a linear order. We say that M is *o-minimal* if every definable set of M (in dimension 1) is a finite union of intervals (closed or open). One can prove that this implies that any model of $Th(M)$ is o-minimal.

As mentioned in the introduction, o-minimality is a major area of research within model theory, as it is both a powerful and versatile framework. This condition was isolated by van den Dries [117] and first systematically studied by Pillay and Steinhorn in [91]. The latter paper establishes the fundamental *cell decomposition* theorem which states that definable subsets in any dimension can be decomposed as a finite union of *cells* (cells in dimension 1 are intervals). This implies that o-minimal theories are geometrically tame: pathologies such as space-filling curves do not exist in them. An important landmark in the subject is the proof by Wilkie [123] that $\mathbb{R}_{\text{exp}}$, the reals with a predicate for the exponential function, is o-minimal. Van den Dries [118] (and later Denef and van den Dries [33]) proved that expanding the reals by adding analytic functions restricted to compact sets also yields an o-minimal structure.

THEOREM A.6. *Any o-minimal theory is dp-minimal.*

PROOF. Let $(M; <, \dots)$ be an o-minimal structure. Let $I = (\bar{b}_i : i \in \mathbb{Q})$ be an indiscernible sequence of n-tuples and let $\phi(x; \bar{y})$ be a formula, where x is a single variable. Then by o-minimality, for every $i \in \mathbb{Q}$, $\phi(x; \bar{b}_i)$ is a finite union of intervals. Note that the end points of those intervals are definable over $\bar{b}_i$. By indiscernibility of the sequence $(\bar{b}_i : i \in \mathbb{Q})$, the number of those intervals and their type (open or closed) is constant as i varies and so are the functions sending $\bar{b}_i$ to each of the end-points of those intervals. In other words, there is a number N, $\emptyset$-definable functions $(g_k(\bar{y}) : k < N)$ and a quantifier free formula $\theta(x; \bar{z})$ in the language $\{=, <\}$ such that for each $i \in \mathbb{Q}$, the formula $\phi(x; \bar{b}_i)$ is equivalent to $\theta(x; g_0(\bar{b}_i), \dots, g_{N-1}(\bar{b}_i))$.

Let $a \in M$ be a singleton. If the sequence $(\text{tp}(\bar{b}_i/a) : i \in \mathbb{Q})$ is not constant, it follows from the previous analysis that a cuts the indiscernible sequence $(\text{dcl}(\bar{b}_i) : i \in \mathbb{Q})$ (where *cut* is defined as in the proof of A.2). If $I = (\bar{b}_i : i \in \mathbb{Q})$ and $J = (\bar{d}_i : i \in \mathbb{Q})$ are mutually indiscernible, then the sequences $(\text{dcl}(\bar{b}_i) : i \in \mathbb{Q})$ and $(\text{dcl}(\bar{d}_i) : i \in \mathbb{Q})$ are mutually indiscernible. Therefore a singleton a cannot cut both. This shows that the theory is dp-minimal. ⊣

For more information about o-minimal theories, we refer the reader to van den Dries' book [119] and Wilkie's survey [124].

A number of weakenings of o-minimality have been introduced. Some of them—such as o-minimal open core—have a more geometric or topological flavor and do not imply NIP. However, others do and we now briefly define two of them.

The structure $(M; <, \dots)$ is *weakly o-minimal* ([34], [78]) if in every elementary extension of M, every definable unary set is a finite union of convex sets. A theory is weakly o-minimal if all its models are (unlike in the case of o-minimality it is not sufficient to look only at one model). An example of a weakly o-minimal structure is the Shelah expansion M^{Sh} of an o-minimal structure M. In this case, the theory of M^{Sh} is weakly o-minimal.

Any weakly o-minimal *theory* is NIP and even dp-minimal. This can be seen by adapting the proof of Theorem A.6.

Going even further, one says that a structure $(M; <, \dots)$ is *quasi o-minimal* ([16]) if any definable unary set is a finite Boolean combination of convex sets and $\emptyset$-definable sets. For example $(\mathbb{Z}; <, 0, 1, +)$ is quasi o-minimal. Again, an adaptation of the previous theorem shows that if a theory T is quasi o-minimal (i.e., all of its models are), then it is dp-minimal.

Finally, we point out that dp-minimal ordered structures constitute an interesting class to study in its own right. This has been done for example in [45] and [112].

A.1.4. C-minimality. The property of C-minimality is to the set of maximal branches of a tree what o-minimality is to linear orders.

Let $(M; \leq)$ be a meet-tree with no leaf, and let B_M be the set of maximal branches of M, i.e., the set of maximal totally ordered subsets of M. Given $a \in B_M$ and $m \in M$, we write $m < a$ if m is in the branch a. Then $(M \cup \{B_M\}, \leq)$ is a meet-tree, the leaves of which are exactly the elements of B_M.

If $a, b, c \in B_M$, we write $C(a, b, c)$ if either $b = c \neq a$ or a, b, c are distinct and $a \wedge c < b \wedge c$. Then the structure (B_M, C) satisfies the following axioms:

(C1) $\forall x, y, z[C(x; y, z) \to C(x, z, y)]$;

(C2) $\forall x, y, z[C(x, y, z) \to \neg C(y, x, z)]$;

(C3) $\forall x, y, z, w[C(x, y, z) \to (C(w, y, z) \vee C(x, w, z)]$;

(C4) $\forall x, y[x \neq y \to \exists z \neq y C(x, y, z)]$.

Conversely, one can show that given a structure $(M; C)$ where $C(x, y, z)$ satisfies axioms (C1)-(C4), then there is a meet-tree $(T; \leq)$ for which M is a subset of the set of branches of T and the C-structure on M coincides with the one defined above.

A structure $(M; C, \dots)$ where (C1)-(C4) are satisfied is called a C-structure.

A C-structure $(M; C, \dots)$ is C-minimal if any definable unary subset is quantifier-free definable using only C and equality. A theory is C-minimal if all of its models are C-minimal.

The main example of a C-minimal structure is that of algebraically closed valued fields (see below).

THEOREM A.7. *Any C-minimal theory is dp-minimal, hence NIP.*

The proof is similar to that of Theorem A.6 using the fact that the formula $C(x; yz)$ has VC-dimension 1.

C-minimality was defined by Macpherson and Steinhorn [79]; a cell decomposition result was shown by Haskell and Macpherson [53]. Further work has been done by Delon, Maalouf, Simonetta.

Both o-minimality and C-minimality are subsumed in the notion of VC-minimal theory defined by Adler in [3] and further studied in [48] and [31].

A.2. Valued fields

Ordered abelian groups. An ordered abelian group is a structure $(\Gamma; 0, +, <)$ such that $(\Gamma; 0, +)$ is an abelian group on which $<$ defines a linear order with the following compatibility condition:

$$x < y \Longrightarrow x + z < y + z, \text{ for all } x, y, z.$$

A quantifier elimination result for ordered abelian groups is proved by Schmitt in [99]. A similar result is proved by Cluckers and Halupczok in [29]. Both languages are somewhat complicated, so we do not give the details here. We however mention two important special cases.

1. Let T_{doag} be the theory of (non-trivial) *divisible* ordered abelian groups (i.e., we add to the theory of ordered abelian groups the axiom $(\exists x) x \neq 0$ and for every integer n, the axiom $(\forall x)(\exists y) n \cdot y = x$.) Then T_{doag} is a complete theory and admits elimination of quantifiers in the language $\{0, +, <\}$.
2. Let T_{pres} be Presburger arithmetic, namely $Th(\mathbb{Z}; 0, 1, +, <)$. That theory does not admit elimination of quantifiers, but it does if we add predicates $\{P_n : n < \omega\}$ defined by $P_n(x) \leftrightarrow (\exists y) n \cdot y = x$.

Using the general quantifier elimination result, Gurevich and Schmitt prove in [50] the following theorem.

THEOREM A.8. *Any ordered abelian group* $(\Gamma; 0, +, <)$ *is NIP.*

It is not true however that any ordered abelian group is dp-minimal, or even strongly dependent.

Example A.9 ([102]). Order $\mathbb{Z}[X]$ by setting $\sum a_i X^i < \sum b_i X^i$ if $a_k < b_k$, where k is maximum such that $a_k \neq b_k$. This makes $(\mathbb{Z}[X]; 0, +, <)$ into an ordered abelian group.

We show that even the pure group $(\mathbb{Z}[X]; 0, +)$ is not strongly dependent.

Let $(p_n : n < \omega)$ list the prime numbers. Define sequences $I_k = (a_i^k < \omega)$, $k < \omega$ by $a_i^k = X^i$ and formulas $\phi_k(x; y)$ saying that $y - x$ is divisible by p_k. For every path $\eta : \omega \to \omega$ one can find, using the Chinese remainder theorem and compactness, a point b_η in the monster model such that $\phi_k(a_l^k; b_\eta)$ holds if and only if $l = \eta(k)$. This gives an ict-pattern of height ω and by Proposition 4.22 shows that $(\mathbb{Z}[X]; 0, +)$ is not strongly dependent.

Valued fields. Since the seminal work of Ax-Kochen and Eršov, valued fields have played an important role in algebraic model theory. They also provide interesting examples of NIP structures.

We will recall briefly some facts about valuations, but we assume some familiarity with this notion in the proofs. The reader is referred to [38] for more details.

Let Γ be an ordered abelian group. We let ∞ be an additional formal element and extend the ordering and composition law on $\Gamma \cup \{\infty\}$ by declaring that ∞ is greater than all elements of Γ and setting $\gamma + \infty = \infty + \gamma = \infty + \infty = \infty$ for all $\gamma \in \Gamma$.

A valued field with value group Γ is a field K equipped with a surjective map $v : K \to \Gamma \cup \{\infty\}$ satisfying:

- $v(x) = \infty \iff x = 0$;
- $v(x + y) \geq \min(v(x), v(y))$;
- $v(xy) = v(x) + v(y)$.

We write K_v to denote the field K equipped with the valuation v.

The *valuation ring* $\mathcal{O}$ of K_v is the ring $\{x \in K : v(x) \geq 0\}$. It is a local ring, i.e., has a unique maximal ideal $\mathfrak{M} = \{x \in K : v(x) > 0\}$. The quotient $\boldsymbol{k} = \mathcal{O}/\mathfrak{M}$ is called the *residue field* of K_v. We let $\text{res} : \mathcal{O} \to \boldsymbol{k}$ be the canonical projection (called the *residue map*).

A valued field can be considered in various languages. The language L_{div} is a one sorted language $(K; 0, 1, +, -, \cdot, |)$ containing the ring language on K and a binary predicate $|$ interpreted as $x|y \iff v(x) \leq v(y)$.

One can also consider valued fields in the three sorted language L_{res} having as sorts K and $\boldsymbol{k}$ equipped with their respective ring structures and Γ equipped with its ordered abelian group structure. We also have two function symbols between sorts: $v : K^* \to \Gamma$ and $\text{res} : \mathcal{O} \to \boldsymbol{k}$ interpreted as the valuation and residue map. (If the reader is bothered by the fact that those functions are not defined everywhere, she can extend them formally by setting $v(0) = 0$ and $\text{res}(x) = 0$ when $v(x) < 0$.)

One can define on any valued field a natural C-structure by

$$C(x, y, z) \equiv v(x - z) < v(y - z).$$

ACVF. Let ACVF denote the theory of algebraically closed non-trivially valued fields in the language L_{div}. The fact that the valued field K_v is algebraically closed forces the residue field $\boldsymbol{k}$ to be also algebraically closed and the value group Γ to be a divisible group.

We let $\mathsf{ACVF}_{(0,0)}$, $\mathsf{ACVF}_{(0,p)}$ and $\mathsf{ACVF}_{(p,p)}$ denote the theories of algebraically closed valued fields where the pair (characteristic of K_v, characteristic of $\boldsymbol{k}$) is respectively equal to $(0, 0)$, $(0, p)$ and (p, p).

THEOREM A.10. *The theory* ACVF *eliminates quantifiers in the language* L_{div}. *Its completion are the theories* $\mathsf{ACVF}_{(0,0)}$, $\mathsf{ACVF}_{(0,p)}$ *and* $\mathsf{ACVF}_{(p,p)}$, *for p a prime number.*

This was essentially proved by Robinson in [97]. See also Chatzidakis [22]. The following easily follows from the theorem.

THEOREM A.11. *The theory* ACVF *equipped with its natural C-structure is C-minimal, in particular it is dp-minimal, hence NIP.*

Henselian valued fields. Recall that a valued field K_v is called *Henselian* if for every polynomial $f(X) \in \mathcal{O}[X]$ and $a \in \mathcal{O}$ such that $\operatorname{res} f(a) = 0$ and $\operatorname{res} f'(a) \neq 0$, there is some $b \in \mathcal{O}$ such that $f(b) = 0$ and $\operatorname{res} a = \operatorname{res} b$.

In particular, any algebraically closed valued field is Henselian. Note that Henselianity is expressible by first order sentences.

An *angular component* is a map $\mathrm{ac} : K_v \to \boldsymbol{k}$ which satisfies:

- $\mathrm{ac}(0) = 0$;
- the restriction of ac to K_v^* has image in $\boldsymbol{k}^*$ and is a morphism of multiplicative groups;
- for any $x \in K_v$ of valuation 0, $\mathrm{ac}(x) = \mathrm{res}(x)$.

If $x, y \in K$, then $v(x) < v(y)$ implies $v(x+y) = v(x)$ and $\mathrm{ac}(x+y) = \mathrm{ac}(x)$. If $v(x) = v(y)$, then either $\mathrm{ac}(x) \neq -\mathrm{ac}(y)$, in which case $v(x+y) = v(x)$ and $\mathrm{ac}(x+y) = \mathrm{ac}(x) + \mathrm{ac}(y)$, or $\mathrm{ac}(x) = -\mathrm{ac}(y)$ in which case $v(x+y) > v(x)$ and we cannot say anything about $\mathrm{ac}(x+y)$.

FACT A.12. *If K_v is an ω_1-saturated valued field, then K_v admits an angular component.*

See for example [23]. Keeping this fact in mind, we may restrict our study to valued fields with an angular component. Let L_{Pas} be the language $L_{\mathrm{res}} \cup \{\mathrm{ac}\}$ where ac is a new function symbol from K to $\boldsymbol{k}$.

THEOREM A.13 (Pas [87]). *The (incomplete) theory of Henselian valued fields of residue characteristic* 0 *eliminates fields quantifiers in the language L_{Pas}.*

See also [22] for a proof. Examples of such fields include $\mathbb{C}((t))$, $\mathbb{R}((t))$: the fields of Laurent series respectively over $\mathbb{R}$, or $\mathbb{C}$, or more generally any field of the type $k((\Gamma))$ where k is a field of characteristic 0 and Γ an ordered abelian group (see [38, Exercise 3.5.6]).

A direct consequence is the celebrated result of Ax-Kochen and Eršov (independently) on elementary equivalence of Henselian valued fields.

THEOREM A.14 (Ax-Kochen, Eršov). *Let K_v and L_w be Henselian valued fields. Denote by Γ_K, Γ_L their respective value groups equipped with the ordered group structure and by $\boldsymbol{k}, \boldsymbol{l}$ the residue fields in the ring language. Assume that $\boldsymbol{k}$ and $\boldsymbol{l}$ have characteristic zero. Then we have:*

$$K_v \equiv L_w \iff (\Gamma_K \equiv \Gamma_L \text{ and } \boldsymbol{k} \equiv \boldsymbol{l}).$$

In the same spirit, if K_v is independent, then this can be traced down to the residue field and the value group.

THEOREM A.15 (Delon [32]). *A Henselian field of residue characteristic* 0 *is NIP if and only if both its value group and its residue field are NIP.*

Knowing Theorem A.8, we do not need to mention the value group.

COROLLARY A.16. *A Henselian field of residue characteristic* 0 *is NIP if and only if its residue field is NIP.*

Delon's proof uses a coheir-counting argument and requires to first understand types. We sketch here a more direct argument using indiscernible sequences. As discussed previously, we may assume that our field K_v is equipped with an angular component ac, and we work in the language L_{Pas}. (The reduct of an NIP theory is NIP, so the result will follow for the languages L_{div} and L_{res}.)

LEMMA A.17. *Let $P(X) = \Sigma_{k\leq n} a_k X^k$ be in $K_v[X]$ and let $(x_i : i < \omega)$ be a sequence of elements of K_v such that $(v(x_i))_{i<\omega}$ is monotonic (increasing or decreasing), then there are $r \leq n$ and $t < \omega$ such that for all $i \geq t$ and $k \neq r$, $v(P(x_i)) = v(a_r x_i^r) < v(a_k x_i^k)$ (hence also $\mathrm{ac}(P(x_i)) = \mathrm{ac}(a_r x_i^r)$).*

PROOF. As the sequence $(v(x_i))_{i<\omega}$ is increasing, there is $t < \omega$ such that for $i \geq t$, $v(x_i)$ has the same relative position with respect to all values of the form $\frac{v(a_k)-v(a_{k'})}{k'-k}$, for $k, k' \leq n$ (if Γ is not divisible, these elements live in its divisible hull, to which the order extends in a unique way). In particular, $v(x_i)$ is not equal to any of these values.

Hence for $i \geq t$, the elements $v(a_k) + k \cdot v(x_i)$ for $k \leq n$ are pairwise distinct and there is $r \leq n$ not depending on i such that their minimum is $v(a_r) + r \cdot v(x_i)$. ⊣

We now classify indiscernible sequences of singletons depending on their type in the C-structure. Let $(x_i : i < \omega)$ be an indiscernible sequence of singletons of K_v.

Case 0: the sequence $(v(x_i) : i < \omega)$ is non-constant. Then by indiscernibility, it is either decreasing or increasing.

We now assume that the sequence $(v(x_i) : i < \omega)$ is constant. For $0 < i < \omega$, let $y_i = x_i - x_0$. Note that it is not possible for the sequence $(v(y_i) : 0 < i < \omega)$ to be increasing. For then, we would have for $i > 1$, $v(x_i - x_1) = v(y_i - y_1) = v(y_1)$. Thus the sequence $v(x_i - x_1)$ would be constant, while the sequence $v(x_i - x_0)$ is not and this contradicts indiscernibility.

Case I: The sequence $(v(y_i) : 0 < i < \omega)$ is decreasing.

Case II: The sequence $(v(y_i) : 0 < i < \omega)$ is constant and the sequence $(\mathrm{ac}(y_i) : 0 < i < \omega)$ is not constant.

Case III: The sequences $(v(y_i) : 0 < i < \omega)$ and $(\mathrm{ac}(y_i) : 0 < i < \omega)$ are both constant. Then we have $v(x_2 - x_1) = v(y_2 - y_1) > v(x_2 - x_0)$. Let x_ω be such that $(x_i : i \leq \omega)$ is indiscernible. For $i < \omega$, we let $z_i = x_i - x_\omega$. Then the sequence $(v(z_i) : i < \omega)$ is increasing.

LEMMA A.18. *Let $(x_i : i < \omega)$ be an indiscernible sequence of singletons of K_v. Then there are:*

- *an indiscernible sequence $(\alpha_i : i < \omega)$ of elements of Γ,*
- *an indiscernible sequence $(b_i : i < \omega)$ of elements of $\mathbf{k}$, such that:*

for any $P(X) \in K_v[X]$, *there are* $r < \omega$ *and* $\gamma \in \Gamma$ *such that* $v(P(x_i)) = \gamma + r \cdot \alpha_i$ *for all* i *large enough.*

Also, there is $q \in \boldsymbol{k}[X]$ *such that for all* i *large enough,* $\mathrm{ac}(P(x_i)) = q(b_i)$.

PROOF. If the sequence $(x_i)_{i<\omega}$ falls in case 0, we are done by Lemma A.17. Assume it falls in case I. There is a polynomial $DP(Y) \in K_v[Y]$ such that we have $P(x_0 + Y) = P(x_0) + DP(Y)$. Then for all $0 < i < \omega$, we have $P(x_i) = P(x_0) + DP(y_i)$. We conclude by applying Lemma A.17 to the sequence $(y_i)_{0<i<\omega}$ and the polynomial $P(x_0) + DP(Y)$.

If $(x_i)_{i<\omega}$ falls in case III, we write similarly $P(x_\omega + Z) = P(x_\omega) + DP_1(Z)$ and apply Lemma A.17 to the sequence $(z_i)_{i<\omega}$ and the polynomial $DP_1(Z)$.

Finally, assume that we are in case II. We then write $P(x_0) + DP(Y) = \sum_{k<n} a_k Y^k$. Let $v_0 = v(y_0)$. Let $A \subseteq n$ be the set of $k < n$ for which $v(a_k) + k \cdot v_0$ is minimal. Let $q(t) \in \boldsymbol{k}[t]$ be the polynomial $\sum_{k\in A} \mathrm{ac}(a_k)t^k$. Then for some $i_* < \omega$, for every $i_* < i < \omega$, $\mathrm{ac}(y_i)$ is not a root of $q(t)$. For such an i, we have $v(P(x_i)) = v(P(x_0) + DP(y_i)) = v(a_k) + k \cdot v_0$, $k \in A$, and $\mathrm{ac}(P(x_i)) = \mathrm{ac}(P(x_0) + DP(y_i)) = q(\mathrm{ac}(y_i))$. The lemma follows. ⊣

We now prove Theorem A.15.

By Theorem A.13 (and keeping Proposition 2.11 in mind), it is enough to show that the following formulas are NIP:

- $\phi(x, \bar{y}) = 0$, where $x, \bar{y}$ are variables of sort K_v and ϕ is a quantifier free formula in the ring language of K_v;
- $\psi(x, \bar{t}(\bar{y}))$, where x is a variable of sort Γ, $\bar{y}$ variables of sorts K_v and Γ, ψ is a formula in the language of Γ, and $\bar{t}$ is a tuple of terms with image in Γ;
- $\theta(x, \bar{t}(\bar{y}))$, where x is a variables of the sort $\boldsymbol{k}$, $\bar{y}$ variables from K_v and $\boldsymbol{k}$, θ is a formula in the language of $\boldsymbol{k}$, and $\bar{t}$ is a tuple of terms with image in $\boldsymbol{k}$;
- $\psi(v(P_1(x, \bar{y}_1)), \ldots, v(P_n(x, \bar{y}_1)), \bar{y}_2)$, where $\bar{y}_2$ are variables from Γ and ψ is a formula in the language of Γ;
- $\theta(\mathrm{ac}(P_1(x, \bar{y}_1)), \ldots, \mathrm{ac}(P_n(x, \bar{y}_1)), \bar{y}_2)$, where $\bar{y}_2$ are variables from $\boldsymbol{k}$ and θ is a formula in the language of $\boldsymbol{k}$.

In the first three cases, the results follows from the fact that algebraically closed fields, $Th(\Gamma)$ and $Th(\boldsymbol{k})$ respectively are NIP.

Assume that the formula $\varphi(x; \bar{y}_1, \bar{y}_2) = \psi(v(P_1(x, \bar{y}_1)), \ldots, v(P_n(x, \bar{y}_1)), \bar{y}_2)$ has IP. Then there is an indiscernible sequence $(x_i : i < \omega)$ of singletons and parameters $\bar{b}_1, \bar{b}_2$ such that $\varphi(x_i; \bar{b}_1, \bar{b}_2)$ holds if and only if i is even.

By Lemma A.18, there is an indiscernible sequence $(\alpha_i : i < \omega)$ of elements of Γ, and for each $k \leq n$ there are $r_k < \omega$ and $\gamma_k \in \Gamma$ such that $v(P_k(x_i, \bar{b}_1)) = \gamma_k + r_k \cdot \alpha_i$ for all i large enough. Hence we can replace each $v(P_k(x, \bar{y}_1))$ in the formula ψ by a term in the language of Γ and we obtain a contradiction to the fact that the ordered abelian group Γ is NIP.

We treat similarly the last case. This finishes the proof of Theorem A.15.

The theory ACVF has been intensely studied in the last ten years. Haskell, Hrushovski and Macpherson provided in [51] a description of imaginaries. They showed in [52] how types can be decomposed into an o-minimal component coming from the value group and a stable *quotient*, internal to the residue field. This property is referred to as *metastability*. A measure-theoretic analog of the Ax-Kochen principle is studied by Hrushovski and Kazhdan [58]. More recently, Hrushovski and Loeser [59] have given a model-theoretic construction of Berkovich spaces from rigid geometry.

The p-adics. Let $\mathbb{Q}_p$ denote the usual field of p-adic numbers. The valued field $\mathbb{Q}_p$ is Henselian, but of residue characteristic p, hence the previous results do not apply. Nonetheless its theory is well understood as we explain now.

First we give an axiomatization of $Th(\mathbb{Q}_p)$ in the language L_{res}. In addition to the paper [10] by Ax and Kochen where the result first appeared, we refer the reader to Cherlin [23, II Th. 40].

THEOREM A.19. *Let T_p be the theory in the language L_{res} expressing that:*

- *K is a Henselian valued fields of characteristic zero;*
- *the residue field $\boldsymbol{k}$ is isomorphic to $\mathbb{F}_p$;*
- *the value group Γ is elementarily equivalent to $(\mathbb{Z}; 0, +, <)$;*
- *$v(p)$ is the smallest positive element of Γ.*

Then T_p is a complete theory, and is equal to the theory of $\mathbb{Q}_p$ in the language L_{res}.

It is worth noting that in the field $\mathbb{Q}_p$, the valuation is definable in the pure field structure (namely, the valuation ring is definable by the formula $\phi(x) = \exists t(t^2 = 1 + p^3y^4))$, hence we also obtain an axiomatization of $Th(\mathbb{Q}_p)$ as a pure field.

The valued field $\mathbb{Q}_p$ does not admit elimination of field quantifiers in either L_{div} or L_{res}. To remedy this, we introduce two enriched languages. First, let L_{mac} be the language L_{div} to which we add unary predicates $\{P_n : 0 < n < \omega\}$. The valued field $\mathbb{Q}_p$ is made into an L_{mac}-structure by interpreting P_n as the set of n-th powers.

THEOREM A.20 (Macintyre [77]). *The field $\mathbb{Q}_p$ has elimination of quantifiers in the language L_{mac}.*

Another way to obtain quantifier elimination is by adding angular components as we did in the case of residue characteristic zero. However, we now need to add infinitely many. Let K_v be a valued field of characteristic 0, whose residue field has characteristic p. A family $(\text{ac}_n : n < \omega)$ is a compatible family of angular components if:

- ac_n is a map $K_v \to \mathcal{O}/p^n\mathcal{O}$;
- $\text{ac}_n(0) = 0$; the restriction of ac_n to K_v^* has image in $(\mathcal{O}/p^n\mathcal{O})^*$ and is a morphism of groups;

- let π_n denote the canonical projection $\pi_n : \mathcal{O}/p^{n+1}\mathcal{O} \to \mathcal{O}/p^n\mathcal{O}$, then $\pi_n \circ \mathrm{ac}_{n+1} = \mathrm{ac}_n$.

Let $L_{\mathrm{Pas},\omega}$ be the language $L_{\mathrm{res}} \cup \{\mathrm{ac}_n : n < \omega\}$.

We can explicitely construct a family of compatible angular components on $\mathbb{Q}_p$. Let $x \in \mathbb{Q}_p^*$ and let $\nu = v(x)$. Set $x_0 = x/p^\nu$. Then $x_0 \in \mathcal{O} \setminus p\mathcal{O}$. Now let $\mathrm{ac}_n(x)$ be the image of x_0 in $\mathcal{O}/p^n\mathcal{O}$.

The properties of the sequence $(\mathrm{ac}_n : 0 < n < \omega)$ are easy to check.

THEOREM A.21 (Belair [15]). *The canonical expansion of $\mathbb{Q}_p$ in the language $L_{\mathrm{Pas},\omega}$ admits elimination of quantifiers.*

Now one can adapt the proof of the previous subsection to show that $\mathbb{Q}_p$ is NIP (or alternatively adapt Delon's proof, which is what Belair does in [15]). Actually, more is true.

THEOREM A.22 (Dolich, Goodrick, Lippel [35]). *For any prime p, the field $\mathbb{Q}_p$ of p-adics in the language L_{div} is dp-minimal.*

Appendix B

PROBABILITY THEORY

We recall here some basic results of probability theory that we need in this text. We refer the reader to any introductory book on the subject for more details. Alon and Spencer's book [7], although not a textbook on probability theory, is a nice reference.

A *probability space* $(\Omega, \mathcal{B}, \mu)$ is a set Ω equipped with a σ-algebra $\mathcal{B}$ and a σ-additive measure μ on $\mathcal{B}$ such that $\mu(\Omega) = 1$. For each integer k, the cartesian power Ω^k is naturally equipped with the product σ-algebra $\mathcal{B}^{\otimes k}$ which is the σ-algebra generated by sets of the form $B_1 \times \cdots \times B_k$ for $B_1, \ldots, B_k \in \mathcal{B}$. The product measure μ^k is defined as the unique probability measure on $(\Omega^k, \mathcal{B}^{\otimes k})$ such that $\mu^k(B_1 \times \cdots \times B_k) = \mu(B_1) \cdots \mu(B_k)$.

A measurable subset $A \subseteq \Omega$ is called an *event*. If A is an event, we let $\mathbf{1}_A$ be its characteristic function. We write $\mathrm{Prob}(A) = \mu(A)$. If $f, g : \Omega \to \mathbb{R}$ are measurable functions, we will write $\{f \geq g\}$ for the event $\{\omega \in \Omega : f(\omega) \geq g(\omega)\}$ and $\mathrm{Prob}(f \geq g)$ instead of $\mathrm{Prob}(\{\omega \in \Omega : f(\omega) \geq g(\omega)\})$.

A measurable function $f : \Omega \to \mathbb{R}$ is called a *random variable*. The probability distribution of f is the probability measure on $\mathbb{R}$ obtained by taking the pushforward of μ by f. It is a common practice to construct a random variable f by specifying only its probability distribution and assume that there is some underlying probability space Ω on which f is defined.

The *expectation* $\mathrm{E}(f)$ of the random variable f is defined as $\int_\Omega f(\omega) d\mu$. Notice that expectation is linear in f. The *variance* of f is $\mathrm{Var}(f) = \mathrm{E}((f - \mathrm{E}(f))^2)$.

More generally, if X is a Borel space, then a *random element* of X is a measurable function $x : \Omega \to X$. The distribution of x is defined as above.

Random variables $f_1, \ldots, f_n : \Omega \to \mathbb{R}$ are said to be *mutually independent* if for any Borel sets $B_1, \ldots, B_n$ of $\mathbb{R}$, we have $\mathrm{Prob}(\bigcap_{k \leq n} \{f_k \in B_k\}) = \prod_{k \leq n} \mathrm{Prob}(f_k \in B_k)$. If $f_1, \ldots, f_n$ are mutually independent, then $\mathrm{E}(f_1 \cdots f_n) = \mathrm{E}(f_1) \cdots \mathrm{E}(f_n)$ and $\mathrm{Var}(f_1 + \cdots + f_n) = \mathrm{Var}(f_1) + \cdots + \mathrm{Var}(f_n)$.

LEMMA B.1 (Union bound). *Let A, B be any two events, then*

$$\mathrm{Prob}(A \cup B) \leq \mathrm{Prob}(A) + \mathrm{Prob}(B).$$

LEMMA B.2 (First moment). *Let $f : \Omega \to \mathbb{R}^+$ be a random variable. Then for all $r > 0$, we have*

$$\mathrm{Prob}(f \geq r) \leq \frac{\mathrm{E}(f)}{r}.$$

PROOF. Write that $\mathrm{E}(f) \geq \mathrm{E}(f \cdot \boldsymbol{1}_{\{f \geq r\}}) \geq r\,\mathrm{Prob}(f \geq r)$. ⊣

PROPOSITION B.3 (Chebyshev's inequality). *Let $f : \Omega \to \mathbb{R}$ be measurable. Then we have*

$$\mathrm{Prob}(|f - \mathrm{E}(f)| \geq \varepsilon) \leq \frac{\mathrm{Var}(f)}{\varepsilon^2}.$$

PROOF. Apply the first moment inequality to the random variable $(f - \mathrm{E}(f))^2$ and $r = \varepsilon^2$. ⊣

PROPOSITION B.4 (Weak law of large numbers). *Let $A \subseteq \Omega$ be an event and fix $\varepsilon > 0$, then for any integer n*

$$\mu^n\left(\bar{\omega} \in \Omega^n : \left|\frac{1}{n}\sum_{i=1}^{n} \boldsymbol{1}_A(\omega_i) - \mu(A)\right| \geq \varepsilon\right) \leq \frac{1}{4n\varepsilon^2}.$$

PROOF. Fix some integer n. For $i \leq n$, the random variable $\boldsymbol{1}_A(\omega_i) : \Omega^n \to \mathbb{R}$ has expectation equal to $\mu(A)$ and variance $\mu(A)(1 - \mu(A)) \leq 1/4$. Also the variables $\boldsymbol{1}_A(\omega_i)$, $i = 1, \dots, n$, are mutually independent. Hence the random variable $\frac{1}{n}\sum_{i=1}^{n} \boldsymbol{1}_A(\omega_i)$ has expectation $\mu(A)$ and variance $\leq \frac{1}{4n}$.

The result then follows from Chebychev's inequality. ⊣

THEOREM B.5 (Chernoff's bound, special case). *Let $f_1, \dots, f_n$ be independent random variables, such that* $\mathrm{Prob}(f_k = 1) = \mathrm{Prob}(f_k = -1) = 1/2$, *for all k. Let $\varepsilon > 0$, then letting $g = \frac{1}{n}\sum f_k$, we have*

$$\mathrm{Prob}(|g| \geq \varepsilon) \leq 2\exp\left(-\frac{n\varepsilon^2}{2}\right).$$

For a proof, see [7, Appendix A].

In the course of the proof of Theorem 6.6, we will need to apply this theorem to a family of random variables f_i, some satisfying the required hypothesis and the others being uniformly equal to zero. The inequality of course also holds in this case. This can be seen by a short calculation: assume that $f_1, \dots, f_m$ are as in the theorem and $f_{m+1}, \dots, f_n$ are equal to zero. Let $g = \frac{1}{n}\sum_{k \leq n} f_k$ and $g' = \frac{1}{m}\sum_{k \leq m} f_k$. We have

$$\mathrm{Prob}(|g| \geq \varepsilon) = \mathrm{Prob}\left(|g'| \geq \frac{n}{m}\varepsilon\right) \leq 2\exp\left(-\frac{m\left(\frac{n\varepsilon}{m}\right)^2}{2}\right) \leq 2\exp\left(-\frac{n\varepsilon^2}{2}\right).$$

REFERENCES

[1] Hans Adler, *Kim's Lemma in NTP2 theories*, handwritten notes available on author's webpage: www.logic.univie.ac.at/~adler/.

[2] ———, *Strong theories, burden, and weight*, draft, available on the author's webpage.

[3] ———, *Theories controlled by formulas of Vapnik-Chervonenkis codimension* 1, draft, available on the author's webpage.

[4] ———, *An introduction to theories without the independence property*, ***Archive of Mathematical Logic***, (2008), to appear.

[5] Noga Alon, *On the density of sets of vectors*, ***Discrete Mathematics***, vol. 46 (1983), no. 2, pp. 199–202.

[6] Noga Alon and Daniel J. Kleitman, *Piercing convex sets and the Hadwiger-Debrunner* (p, q) *problem*, ***Advances in Mathematics***, vol. 96 (1992), pp. 103–112.

[7] Noga Alon and Joel H. Spencer, ***The Probabilistic Method***, Wiley-Interscience series in discrete mathematics and optimization, Wiley, 2000.

[8] Matthias Aschenbrenner, Alf Dolich, Deirdre Haskell, Dugald Macpherson, and Sergei Starchenko, *Vapnik-Chervonenkis density in some theories without the independence property, I*, to appear in the Transactions of the American Mathematical Society, 2013.

[9] ———, *Vapnik-Chervonenkis density in some theories without the independence property, II*, ***Notre Dame Journal of Formal Logic***, vol. 54 (2013), no. 3–4, pp. 311–363.

[10] James Ax and Simon Kochen, *Diophantine problems over local fields II. a complete set of axioms for p-adic number theory*, ***American Journal of Mathematics***, vol. 87 (1965), no. 3, pp. 631–648.

[11] Yerzhan Baisalov and Bruno Poizat, *Paires de structures o-minimales*, ***The Journal of Symbolic Logic***, vol. 63 (1998), pp. 570–578.

[12] John T. Baldwin, ***Fundamentals of Stability Theory***, Perspectives in Mathematical Logic, Springer-Verlag, 1988.

[13] John T. Baldwin and Michael Benedikt, *Stability theory, permutations of indiscernibles, and embedded finite models*, ***Transactions of the American Mathematical Society***, vol. 352 (2000), no. 11, pp. 4937–4969.

[14] JOHN T. BALDWIN and JAN SAXL, *Logical stability in group theory*, ***Journal of the Australian Mathematical Society (Series A)***, vol. 21 (1976), no. 3, pp. 267–276.

[15] LUC BÉLAIR, *Types dans les corps valués munis d'applications coefficients*, ***Illinois Journal of Mathematics***, vol. 43 (1999), no. 2, pp. 410–425.

[16] OLEG BELEGRADEK, YA'ACOV PETERZIL, and FRANK WAGNER, *Quasi-o-minimal structures*, ***The Journal of Symbolic Logic***, vol. 65 (2000), no. 3, pp. 1115–1132.

[17] ITAÏ BEN YAACOV, *Continuous and random Vapnik-Chervonenkis classes*, ***Israel Journal of Mathematics***, vol. 173 (2009), pp. 309–333.

[18] ITAÏ BEN YAACOV and ARTEM CHERNIKOV, *An independence theorem for NTP2 theories*, to appear in the Journal of Symbolic Logic.

[19] ALEXANDER BERENSTEIN, ALFRED DOLICH, and ALF ONSHUUS, *The independence property in generalized dense pairs of structures*, ***The Journal of Symbolic Logic***, vol. 76 (2011), no. 2, pp. 391–404.

[20] GARETH BOXALL, *NIP for some pair-like theories*, ***Archive for Mathematical Logic***, vol. 50 (2011), pp. 353–359.

[21] ENRIQUE CASANOVAS, ***Simple Theories and Hyperimaginaries***, Lecture Notes in Logic, Cambridge University Press, 2011.

[22] ZOÉ CHATZIDAKIS, *Théorie des modèles des corps valués*, http://www.logique.jussieu.fr/~zoe/.

[23] GREGORY L. CHERLIN, ***Model-Theoretic Algebra: Selected Topics***, Lecture Notes in Mathematics, vol. 52, Springer-Verlag, Berlin, 1976.

[24] ARTEM CHERNIKOV, *Theories without the tree property of the second kind*, ***Annals of Pure and Applied Logic***, vol. 165 (2014), pp. 695–723.

[25] ARTEM CHERNIKOV and ITAY KAPLAN, *Forking and dividing in* NTP_2 *theories*, ***The Journal of Symbolic Logic***, vol. 77 (2012), no. 1, pp. 1–20.

[26] ARTEM CHERNIKOV, ANAND PILLAY, and PIERRE SIMON, *External definability and groups in NIP theories*, to appear in the Journal of the London Mathematical Society.

[27] ARTEM CHERNIKOV and PIERRE SIMON, *Externally definable sets and dependent pairs II*, to appear in the Transactions of the American Mathematical Society.

[28] ———, *Externally definable sets and dependent pairs*, ***Israel Journal of Mathematics***, vol. 194 (2013), no. 1, pp. 409–425.

[29] RAF CLUCKERS and IMMANUEL HALUPCZOK, *Quantifier elimination in ordered abelian groups*, ***Confluentes Mathematici***, vol. 3 (2011), pp. 587–615.

[30] ANNALISA CONVERSANO and ANAND PILLAY, *Connected components of definable groups and o-minimality I*, ***Advances in Mathematics***, vol. 231 (2012), pp. 605–623.

[31] SARAH COTTER and SERGEI STARCHENKO, *Forking in VC-minimal theories*, ***The Journal of Symbolic Logic***, vol. 77 (2012), no. 4, pp. 1257–1271.

[32] FRANÇOISE DELON, *Types sur $C((X))$*, ***Study Group on Stable Theories, Second year: 1978/79***, Secrétariat Math., 1981.

[33] JAN DENEF and LOU VAN DEN DRIES, *p-adic and real subanalytic sets*, ***The Annals of Mathematics***, vol. 128 (1988), no. 1, pp. 79–138.

[34] MAX DICKMANN, *Elimination of quantifiers for ordered valuation rings*, ***The Journal of Symbolic Logic***, vol. 52 (1987), no. 1, pp. 116–128.

[35] ALFRED DOLICH, DAVID LIPPEL, and JOHN GOODRICK, *Dp-minimal theories: basic facts and examples*, ***Notre Dame Journal of Formal Logic***, vol. 52 (2011), no. 3, pp. 267–288.

[36] CLIFTON EALY, KRZYSZTOF KRUPIŃSKI, and ANAND PILLAY, *Superrosy dependent groups having finitely satisfiable generics*, ***Annals of Pure and Applied Logic***, vol. 151 (2008), pp. 1–21.

[37] CLIFTON EALY and ALF ONSHUUS, *Characterizing rosy theories*, ***The Journal of Symbolic Logic***, vol. 72 (2007), no. 3, pp. 919–940.

[38] A. J. ENGLER and A. PRESTEL, ***Valued Fields***, Springer Monographs in Mathematics, Springer, 2006.

[39] JOSEPH FLENNER and VINCENT GUINGONA, *Convexly orderable groups and valued fields*, to appear in the Journal of Symbolic Logic.

[40] SALLY FLOYD and MANFRED WARMUTH, *Sample compression, learnability, and the Vapnik-Chervonenkis dimension*, ***Machine Learning***, 1995, pp. 269–304.

[41] PETER FRANKL, *On the trace of finite sets*, ***Journal of Combinatorial Theory, Series A***, vol. 34 (1983), no. 1, pp. 41–45.

[42] JAKUB GISMATULLIN, *Model theoretic connected components of groups*, ***Israel Journal of Mathematics***, vol. 184 (2011), pp. 251–274.

[43] JAKUB GISMATULLIN and KRZYSZTOF KRUPIŃSKI, *On model-theoretic connected components in some group extensions*, preprint.

[44] JAKUB GISMATULLIN, DAVIDE PENAZZI, and ANAND PILLAY, *Some model theory of $SL(2,R)$*, preprint.

[45] JOHN GOODRICK, *A monotonicity theorem for dp-minimal densely ordered groups*, ***The Journal of Symbolic Logic***, vol. 75 (2010), no. 1, pp. 221–238.

[46] VINCENT GUINGONA, *Dependence and isolated extensions*, ***Proceedings of the American Mathematical Society***, vol. 139 (2011), pp. 3349–3357.

[47] ———, *On uniform definability of types over finite sets*, ***The Journal of Symbolic Logic***, vol. 77 (2012), no. 2, pp. 499–514.

[48] VINCENT GUINGONA and M. C. LASKOWSKI, *On VC-minimal theories and variants*, ***Archive for Mathematical Logic***, vol. 52 (2013), no. 7, pp. 743–758.

[49] AYHAN GÜNAYDIN and PHILIPP HIERONYMI, *Dependent pairs*, ***The Journal of Symbolic Logic***, vol. 76 (2011), no. 2, pp. 377–390.

[50] YURI GUREVICH and PETER H. SCHMITT, *The theory of ordered abelian groups does not have the independence property*, ***Transactions of the American Mathematical Society***, vol. 284 (1984), no. 1, pp. 171–182.

[51] Deirdre Haskell, Ehud Hrushovski, and Dugald Macpherson, *Definable sets in algebraically closed valued fields. Part I: elimination of imaginaries*, ***Journal für die reine und angewandte Mathematik***, vol. 597 (2006), pp. 175–236.

[52] ———, ***Stable Domination and Independence in Algebraically Closed Valued Fields***, Lecture Notes in Logic, vol. 30, Association for Symbolic Logic, Chicago, IL, 2008.

[53] Deirdre Haskell and Dugald Macpherson, *Cell decompositions of C-minimal structures*, ***Annals of Pure and Applied Logic***, vol. 66 (1994), no. 2, pp. 113–162.

[54] Assaf Hasson and Alf Onshuus, *Stable types in rosy theories*, ***The Journal of Symbolic Logic***, vol. 75 (2010), no. 4, pp. 1211–1230.

[55] ———, *Unstable structures definable in o-minimal theories*, ***Selecta Mathematica***, vol. 16 (2010), no. 1, pp. 121–143.

[56] Wilfrid Hodges, ***A Shorter Model Theory***, Cambridge University Press, 1997.

[57] Wilfrid Hodges, A. H. Lachlan, and Saharon Shelah, *Possible orderings of an indiscernible sequence*, ***Bulletin of the London Mathematical Society***, vol. 9 (1977), no. 2, pp. 212–215.

[58] Ehud Hrushovski and David Kazhdan, *Integration in valued fields*, ***Algebraic Geometry and Number Theory*** (Victor Ginzburg, editor), Progress in Mathematics, vol. 253, Birkhäuser Boston, 2006, pp. 261–405.

[59] Ehud Hrushovski and Francois Loeser, *Non-archimedean tame topology and stably dominated types*, preprint.

[60] Ehud Hrushovski, Ya'acov Peterzil, and Anand Pillay, *Groups, measures, and the NIP*, ***Journal of the American Mathematical Society***, vol. 21 (2008), no. 2, pp. 563–596.

[61] Ehud Hrushovski and Anand Pillay, *On NIP and invariant measures*, ***Journal of the European Mathematical Society***, vol. 13 (2011), pp. 1005–1061.

[62] Ehud Hrushovski, Anand Pillay, and Pierre Simon, *A note on generically stable measures and fsg groups*, ***Notre Dame Journal of Formal Logic***, vol. 53 (2012), no. 4, pp. 599–605.

[63] ———, *Generically stable and smooth measures in NIP theories*, ***Transactions of the American Mathematical Society***, vol. 365 (2013), pp. 2341–2366.

[64] Hunter R. Johnson and Micheal C. Laskowski, *Compression schemes, stable definable families, and o-minimal structures*, ***Discrete & Computational Geometry***, vol. 43 (2010), no. 4, pp. 914–926.

[65] Itay Kaplan, Alf Onshuus, and Alex Usvyatsov, *Additivity of the dp-rank*, ***Transactions of the American Mathematical Society***, vol. 365 (2013), pp. 5783–5804.

[66] Itay Kaplan, Thomas Scanlon, and Frank O. Wagner, *Artin-Schreier extensions in NIP and simple fields*, ***Israel Journal of Mathematics***, vol. 185 (2011), no. 1, pp. 141–153.

[67] ITAY KAPLAN and SAHARON SHELAH, *A dependent theory with few indiscernibles*, to appear in the Israel Journal of Mathematics.

[68] ———, *Examples in dependent theories*, to appear in the Journal of Symbolic Logic.

[69] ———, *Chain conditions in dependent groups*, ***Annals of Pure and Applied Logic***, vol. 164 (2013), no. 12, pp. 1322–1337.

[70] ITAY KAPLAN and ALEX USVYATSOV, *Strict independence in dependent theories*, to appear in the Journal of Mathematical Logic.

[71] H. JEROME KEISLER, *Measures and forking*, ***Annals of Pure and Applied Logic***, vol. 34 (1987), no. 2, pp. 119–169.

[72] KRZYSZTOF KRUPIŃSKI, *Fields interpretable in rosy theories*, ***Israel Journal of Mathematics***, vol. 175 (2010), pp. 421–444.

[73] ———, *Fields interpretable in superrosy groups with NIP (the non-solvable case)*, ***The Journal of Symbolic Logic***, vol. 75 (2010), pp. 372–386.

[74] MICHAEL C. LASKOWSKI, *Vapnik-Chervonenkis classes of definable sets*, ***Journal of the London Mathematical Society***, vol. 45 (1992), pp. 377–384.

[75] ROI LIVNI and PIERRE SIMON, *Honest compressions and their application to compression schemes*, ***Proceedings of the 26th Conference on Learning Theory (COLT)***, 2013.

[76] NEWELSKI LUDOMIR, *Topological dynamics of definable group actions*, ***The Journal of Symbolic Logic***, vol. 74 (2009), pp. 50–72.

[77] ANGUS MACINTYRE, *On definable subsets of p-adic fields*, ***The Journal of Symbolic Logic***, vol. 41 (1976), no. 3, pp. 605–610.

[78] DUGALD MACPHERSON, DAVID MARKER, and CHARLES STEINHORN, *Weakly o-minimal structures and real closed fields*, ***Transactions of the American Mathematical Society***, vol. 352 (2000), no. 12, pp. 5435–5483.

[79] DUGALD MACPHERSON and CHARLES STEINHORN, *On variants of o-minimality*, ***Annals of Pure and Applied Logic***, vol. 79 (1996), no. 2, pp. 165–209.

[80] DUGALD MACPHERSON and KATRIN TENT, *Pseudofinite groups with NIP theory and definability in finite simple groups*, ***Groups and Model Theory***, Contemp. Math. Series, vol. 576, AMS, 2012.

[81] DAVID MARKER, ***Model Theory: An Introduction***, Springer, 2002.

[82] JIRÍ MATOUSEK, *Bounded VC-dimension implies a fractional Helly theorem*, ***Discrete & Computational Geometry***, vol. 31 (2004), pp. 251–255.

[83] WILLIAM MITCHELL, *Aronszajn trees and the independence of the transfer property*, ***Annals of Mathematical Logic***, vol. 5 (1972), pp. 21–46.

[84] LUDOMIR NEWELSKI, *Model theoretic aspects of the Ellis semigroup*, ***Israel Journal of Mathematics***, vol. 190 (2012), no. 1, pp. 477–507.

[85] ALF ONSHUUS, *Properties and consequences of thorn-independence*, ***The Journal of Symbolic Logic***, vol. 71 (2006), no. 1, pp. 1–21.

[86] ALF ONSHUUS and ALEXANDER USVYATSOV, *On dp-minimality, strong dependence and weight*, ***The Journal of Symbolic Logic***, vol. 76 (2011), no. 3, pp. 737–758.

[87] JOHAN PAS, *Uniform p-adic cell decomposition and local zeta functions*, ***Journal für die reine und angewandte Mathematik***, vol. 399 (1989), pp. 137–172.

[88] YA'ACOV PETERZIL, *Pillay's conjecture and its solution—a survey*, ***Logic Colloquium 2007***, Cambridge University Press, 2010.

[89] ANAND PILLAY, ***Geometric Stability Theory***, Oxford Logic Guides, Clarendon Press, 1996.

[90] ———, *Topological dynamics and definable groups*, ***The Journal of Symbolic Logic***, vol. 78 (2013), no. 2, pp. 657–666.

[91] ANAND PILLAY and CHARLES STEINHORN, *Definable sets in ordered structures I*, ***Transactions of the American Mathematical Society***, vol. 295 (1986), no. 2, pp. 565–592.

[92] ANAND PILLAY and PREDRAG TANOVIĆ, *Generic stability, regularity, and quasi-minimality*, preprint.

[93] BRUNO POIZAT, *Théories instables*, ***The Journal of Symbolic Logic***, vol. 46 (1981), no. 3, pp. 513–522.

[94] ———, *Post-Scriptum à "Théories Instables"*, ***The Journal of Symbolic Logic***, vol. 48 (1983), no. 1, pp. 60–62.

[95] ———, ***Cours de théorie des modèles***, Nur al-Mantiq wal-Mari'fah, 1985, (English translation: ***A Course in Model Theory***, Springer.).

[96] ———, ***Groupes stables***, Nur al-Mantiq wal-Ma'rifah, 1987.

[97] ABRAHAM ROBINSON, ***Complete Theories***, North-Holland, Amsterdam, 1956.

[98] N. SAUER, *On the density of families of sets*, ***Journal of Combinatorial Theory, Series A***, vol. 13 (1972), no. 1, pp. 145–147.

[99] P. H. SCHMITT, ***Model Theory of Ordered Abelian Groups***, Universität Heidelberg, 1982, Habilitationsschrift.

[100] ALEXANDER SCHRIJVER, ***Theory of Linear and Integer Programming***, Wiley, New York, 1986.

[101] SAHARON SHELAH, *Definable groups for dependent and 2-dependent theories*, preprint.

[102] ———, *Strongly dependent theories*, preprint.

[103] ———, *Stability, the f.c.p., and superstability; model theoretic properties of formulas in first order theory*, ***Annals of Mathematical Logic***, vol. 3 (1971), no. 3, pp. 271–362.

[104] ———, *A combinatorial problem; stability and order for models and theories in infinitary languages*, ***Pacific Journal of Mathematics***, vol. 41 (1972), pp. 247–261.

[105] ———, *Simple unstable theories*, ***Annals of Mathematical Logic***, vol. 19 (1980), no. 3, pp. 177–203.

[106] ———, ***Classification Theory and the Number of Nonisomorphic Models***, second ed., Studies in Logic and the Foundations of Mathematics, vol. 92, North-Holland Publishing Co., Amsterdam, 1990.

[107] ———, *Classification theory for elementary classes with the dependence property—a modest beginning*, ***Scientiae Mathematicae Japonicae***, vol. 59 (2004), no. 2, pp. 265–316, special issue on set theory and algebraic model theory.

[108] ———, *Minimal bounded index subgroup for dependent theories*, ***Proceedings of the American Mathematical Society***, vol. 136 (2008), no. 3, pp. 1087–1091.

[109] ———, *Dependent first order theories, continued*, ***Israel Journal of Mathematics***, vol. 173 (2009), no. 1, pp. 1–60.

[110] ———, *A dependent dream and recounting types*, (2010), preprint, 950 in Shelah archive.

[111] ———, *Dependent theories and the generic pair conjecture*, ***Communications in Contemporary Mathematics***, vol. submitted (2010), math.LO/0702292.

[112] PIERRE SIMON, *On dp-minimal ordered structures*, ***The Journal of Symbolic Logic***, vol. 76 (2011), no. 4, pp. 448–460.

[113] ———, *Finding generically stable measures*, ***The Journal of Symbolic Logic***, vol. 77 (2012), no. 1, pp. 263–278.

[114] ———, *Distal and non-distal theories*, ***Annals of Pure and Applied Logic***, vol. 164 (2013), no. 3, pp. 294–318.

[115] KATRIN TENT and MARTIN ZIEGLER, ***A Course in Model Theory***, Lecture Notes in Logic, Cambridge University Press, 2012.

[116] ALEXANDER USVYATSOV, *On generically stable types in dependent theories*, ***The Journal of Symbolic Logic***, vol. 74 (2009), no. 1, pp. 216–250.

[117] LOU VAN DEN DRIES, *Remarks on Tarski's problem concerning* $(R, +, *, exp)$, ***Logic Colloquium '82*** (G. Lolli, G. Longo, and A. Marcja, editors), Studies in Logic and the Foundations of Mathematics, vol. 112, Elsevier, 1984, pp. 97–121.

[118] ———, *A generalization of the Tarski-Seideberg theorem, and some nondefinability results*, ***Bulletin of the American Mathematical Society***, vol. 15 (1986), pp. 189–193.

[119] ———, ***Tame Topology and O-minimal Structures***, London Mathematical Society Lecture Note Series, Cambridge University Press, 1998.

[120] VLADIMIR N. VAPNIK and ALEXEY YA. CHERVONENKIS, *On the uniform convergence of relative frequencies of events to their probabilities*, ***Theory of Probability and its Applications***, vol. 16 (1971), no. 2, pp. 264–280.

[121] FRANK O. WAGNER, ***Stable Groups***, London Mathematical Society Lecture Note Series, Cambridge University Press, 1997.

[122] ———, ***Simple Theories***, Mathematics and its Applications, vol. 503, Kluwer Academic Publishers, Dordrecht, 2000.

[123] ALEX WILKIE, *Model completeness results for expansions of the ordered field of real numbers by restricted Pfaffian functions and the exponential function*, ***Journal of the American Mathematical Society***, vol. 9 (1996), pp. 1051–1094.

[124] ———, *O-minimal structures*, ***Séminaire Bourbaki (2007/08), exp. nº 985, Astérisque***, vol. 326 (2009), pp. 131–142.

INDEX

For EU product safety concerns, contact us at Calle de José Abascal, 56–1°, 28003 Madrid, Spain or eugpsr@cambridge.org.

www.ingramcontent.com/pod-product-compliance
Lightning Source LLC
LaVergne TN
LVHW101503200926
848483LV00001B/208

* 9 7 8 1 1 0 7 0 5 7 7 5 3 *